LEADING WITH AI AGENTS

LEADING WITH AI AGENTS

A LEADER'S PLAYBOOK TO STRATEGIZE, BUILD, AND GOVERN DIGITAL WORKFORCES

REDDY MALLIDI

PUBLISHED BY
reddymallidi.com

LEADING WITH AI AGENTS
A Leader's Playbook to Strategize, Build, and Govern Digital Workforces

ISBN:
979-8-9931029-1-7 (Hardcover)
979-8-9931029-0-0 (Paperback)
979-8-9931029-2-4 (eBook)

For ordering this and other titles, or to download valuable AI resources, please visit: www.reddymallidi.com

To my late aunt, who always believed in me and taught me never to give up.

To my parents, who instilled integrity and the belief that things would always work out.

To my teachers at APR School, Tadikonda, who sparked a lifelong passion for learning.

You shaped the foundation of who I am.

This is for you—with gratitude and love.

Contents

INTRODUCTION

The $150 Million Question

Over the last few years, I have had the privilege of delivering more than $150 million in annual savings through AI implementations across Fortune 2000 companies. From automating export compliance screening for 50 million accounts, reconciling 80 million daily financial transactions to automating suture inspection with less than 1% error rate, I have witnessed AI's transformative power firsthand.

But here's what I have learned: the next wave of AI isn't limited to automation—it centers on intelligent agents that can reason, adapt, and act to achieve business objectives.

Today's executives and professionals face a critical inflection point. You have likely experimented with ChatGPT, tried Copilot, or deployed basic AI tools. Yet the leap from individual AI interactions to enterprise-grade AI agents capable of transforming operations feels daunting. The chasm between AI experimentation and true AI transformation remains wide.

This book bridges that gap.

The Agent Advantage

AI agents represent a fundamental shift from reactive tools to proactive partners. Unlike traditional automation that follows rigid scripts, AI agents analyze complex data, understand context, and make intelligent decisions to achieve specific goals.

Consider MIKI, the AI agent deployed by BMC Software's global sales organization. According to Paul Cant, Chief Revenue Officer, MIKI didn't merely automate routine tasks. It analyzed CRM data, activity patterns, sentiment analysis, and pipeline data across multiple sources to flag deal risks and provide next-best-action recommendations. Acting as a "chief of staff," MIKI helped sales teams shift from spending 70% of their time understanding data to making faster, more accurate decisions. The result? Sales operations evolved from manual, Excel-based forecasting that consumed over 2,000 hours weekly to AI-driven strategic revenue intelligence that delivered multi-million dollar productivity gains and improved forecast accuracy.

Similar scenarios are happening across industries. Organizations that master AI agents gain sustainable competitive advantages: faster decision-making, personalized customer experiences, and strategic focus.

Your Strategic Guide

Leading With AI Agents is designed for C-suite executives, business leaders, and professionals who need to operationalize AI agents within their organizations. Through proven frameworks, case studies, and practical strategies from successful enterprise implementations, this book provides the strategic insight you need to understand what AI agents can accomplish, when they are worth the investment, and how to deploy them effectively.

What You Will Master

The book is structured to take you from strategic understanding to operational readiness:

- **Understanding AI Agent Fundamentals**: We begin by exploring what distinguishes AI agents from traditional automation, examining their core capabilities of perception, reasoning, memory, and action.
- **Architecting Intelligent Systems**: Discover the design patterns and orchestration mechanisms that enable sophisticated agent behavior, from structured workflows to dynamic, self-directing systems.
- **Building Multi-Agent Ecosystems**: Learn how individual agents collaborate to form intelligent networks that can tackle complex business challenges.
- **Designing Human-Agent Partnerships**: Explore how to structure effective collaboration between human teams and AI agents, including interface design, trust protocols, and oversight systems.
- **From Strategy to Implementation**: Follow a systematic approach to building your first AI agent, from defining objectives and selecting frameworks to testing and monitoring.
- **Scaling Digital Workforces**: Master the operational complexities of managing AI agent ecosystems, including infrastructure, cost optimization, and performance monitoring.
- **Driving Business Value**: Examine compelling use cases across functions—from customer service to supply chain—with measurable impact.
- **Building Responsibly**: Navigate the critical foundations of security, safety, and governance that turn AI agents into sustainable competitive advantage.

Why This Matters Now

Early adopters are already transforming their operations, setting new standards for productivity, customer experience, and innovation. The business opportunity is substantial—McKinsey research suggests generative AI could deliver up to $4.4 trillion in annual value globally.[1] Yet most organizations struggle to move beyond pilot projects to scaled implementations.

Organizations that delay risk falling behind competitors. However, rushed implementations can be costly. Poorly designed AI agents can damage customer trust, create operational chaos, or invite regulatory scrutiny.

The challenge extends beyond technical issues—it is strategic and operational. How do you design AI agents that align with business objectives? How do you ensure they integrate seamlessly with existing systems? How do you govern autonomous agents while maintaining control and compliance?

As Mark Cuban observed, "If you are a CEO, you can't just say, I am going to get my tech guys to understand it and educate me. You must understand it because it will have a significant impact on every single thing you do."[2]

Success requires disciplined strategy, proven design patterns, and robust governance frameworks. This book emphasizes the strategic and operational principles that separate successful AI agent deployments from expensive failures. As software engineering evolved from ad-hoc coding to disciplined design patterns, AI agent development must mature through proven methodologies.

Your Next Move

AI agents are reshaping industries now. Companies that act early will define the new landscape.

Whether you are evaluating your first AI agent pilot or scaling an enterprise-wide deployment, this book provides the strategic framework, operational guidance, and governance principles you need to turn AI agents into sustainable business value.

The organizations that master AI agents will set the pace for their industries. Start building your AI advantage today.

1

Automation to Intelligence: The Rise of AI Agents

"AI is the new electricity. Just as one hundred years ago electricity transformed industry after industry, AI will do the same now."

— Andrew Ng

Paul Cant's enterprise sales team at BMC Software, a global technology company with billions in revenue, recognized the fundamental limitations of their fragmented sales technology approach. Despite investing heavily in multiple point solutions, they were still trapped in manual processes. Sales teams were spending over 2,000 hours weekly preparing forecasts using Excel, while managers cobbled together insights from disparate systems.

"We were spending over 2,000 hours a week across the globe in preparing forecasts like that," Paul explained. "Excel at the center of that, despite having a bunch of tools." As a seasoned CRO committed to transformation, he sought a solution that would consolidate their tech stack into a purpose-built AI platform and enable data-driven decision-making at global scale.

Then BMC implemented MIKI, Aviso's AI agent designed for revenue intelligence, as the centerpiece of a comprehensive sales technology transformation. Beyond automating tasks, MIKI analyzed data from CRM systems, activity patterns, sentiment analysis, and pipeline data across multiple sources, identifying patterns and generating actionable insights that no human could synthesize at scale.

MIKI was intelligently coordinating complex organizational data. The agent perceived signals in deal progression and engagement patterns, reasoned about risk assessment based on historical performance, and took action by providing AI-driven forecasts, flagging at-risk opportunities, and delivering next-best-action recommendations. Acting as Paul's "chief of staff," MIKI helped assimilate the vast amounts of information needed to run a billion-dollar revenue operation.

The transformation was dramatic. Paul's team moved from spending 70% of their time trying to understand data to making faster, more effective decisions. As Paul described it: "It's all about decision-making in sales. By nature of having better data—presented in a contextual way—you can say 'Watch out here. This opportunity doesn't look like it's tracking the way it should be based on history.'"

The 15-20% productivity improvements across a global sales organization translated to significant revenue growth without adding capacity. The results spoke for themselves. BMC delivered multi-million dollar ROI over three years through productivity savings and revenue increases, while experiencing their largest growth years ever during Paul's CRO tenure.

MIKI's story exemplifies the core concept of this book: AI agents. These systems leap beyond simple automation, embodying a form of artificial intelligence capable of dynamic interaction and goal-oriented behavior.

Many of us have dealt with chatbots before. But what's new about AI agents?

What Is an AI Agent?

An AI agent is an autonomous or semi-autonomous system that operates within a defined environment.[1] Examples include a robot, a software application, or a network or information stream. At its core, AI agents possess four fundamental capabilities that enable them to function effectively.

Perception: Agents gather information about their environment through sensors, data inputs, or other interface mechanisms, allowing them to understand their current situation and detect changes.

Memory: They maintain internal storage of past experiences, learned knowledge, and current state information. This memory is essential for understanding context, learning from experience, and making informed decisions in observable environments.

Reasoning: Agents process perceived information alongside stored knowledge to make decisions. Doing so involves analyzing current conditions, comparing them with past experiences, and determining the best course of action based on their programming and objectives.

Action: Lastly, they execute decisions through outputs, actuators, or other mechanisms that allow them to influence their environment and work toward achieving their specific goals.

Key Characteristics

Effective AI agents typically demonstrate several important characteristics. These characteristics work together to create systems that can handle complex, dynamic environments while maintaining focus on their intended objectives.

- Autonomy enables agents to operate independently, making decisions and executing tasks without requiring constant human oversight or intervention.
- Reactivity allows agents to perceive environmental changes and respond appropriately to new conditions or stimuli as they arise.
- Proactiveness drives agents beyond reactive responses to take initiative and pursue goal-directed behavior.
- Adaptability provides agents with the capacity to modify their behavior based on experience, improving their performance over time through learning mechanisms.
- Social ability enables agents to interact and communicate effectively with other agents, systems, or human users when collaboration or coordination is required.

AI Agents vs. Bots

Most of us interacted with or heard of Robotic Process Automation bots, commonly known as RPA bots. While both bots and AI agents aim to automate tasks, their underlying capabilities differ dramatically:

RPA Bots are primarily rule-based systems designed to mimic human actions for repetitive, structured tasks within digital systems such as filling out forms and copying data between applications. They follow predefined scripts and typically lack the ability to handle exceptions, learn, or adapt to changing interfaces or unstructured data.

AI Agents utilize AI techniques like machine learning and natural language processing. They can handle unstructured data including text and images, learn from experience, make nuanced decisions, adapt to dynamic environments, and tackle more complex, non-routine tasks.

Historical Evolution

The concept of AI agents evolved alongside the field of Artificial Intelligence itself.

Mid-20th Century: Early AI research explored symbolic reasoning and problem-solving. Concepts like the General Problem Solver laid theoretical groundwork, and early robots like Shakey in late 1960s, demonstrated integrating perception, planning, and action, albeit in highly controlled environments.[2]

1980s: Focus shifted towards knowledge representation and expert systems. Reactive architectures, such as Rodney Brooks' subsumption architecture, emerged, emphasizing direct links between perception and action for robust real-world interaction, bypassing complex symbolic reasoning for simpler behaviors.[3]

1990s: Formal agent theories gained prominence. Models like Belief-Desire-Intention (BDI) provided frameworks for designing rational, deliberative agents capable of planning based on internal states. The rise of the internet spurred interest in software agents for information retrieval and e-commerce.[4]

2000s: Advances in machine learning, principally reinforcement learning, allowed agents to learn optimal behaviors through trial and error. Multi-agent system research explored cooperation and competition among agents.[5]

2010s - Present: The deep learning advancements, fueled by massive datasets and computational power, have dramatically enhanced agent capabilities. Agents can now process text, images, and audio, learn intricate strategies, and power applications ranging from hyper-personalized virtual assistants to highly capable autonomous systems.[6]

This evolution reflects a continuous progression from simple rule-based systems to agents possessing sophisticated reasoning, planning, and learning capabilities, enabling them to tackle increasingly complex problems.

When to Use AI Agents

While powerful, AI agents introduce complexity and overhead compared to simpler scripts or direct API calls. Deploying an agent makes sense primarily when its unique capabilities offer significant advantages. Consider using an agent under these circumstances:

Open-Ended or Complex Problems: When the path to a solution isn't predetermined or involves navigating dynamic conditions, an agent's ability to continuously perceive, reason, and adapt its strategy based on real-time feedback becomes essential. In scenarios where you can't hard-code a fixed sequence of actions or predict the exact number of steps needed, this adaptive capability proves invaluable.

Multiple System Dependency: Tasks often require orchestrating interactions across various systems – querying databases, calling external APIs, searching knowledge bases, running computations. An agent can manage these multi-step workflows, maintain context across interactions, decide which tool to use when, and synthesize information coherently.

Autonomous Operations and Scalability: When processes need to run continuously, handle large volumes, or operate in environments where constant human oversight is impractical or impossible, an autonomous agent becomes essential. Once properly configured and validated, it can execute tasks end-to-end while scaling operations efficiently.

Adaptability: If the environment changes frequently, or if performance needs to improve over time based on experience, a learning agent is often necessary. Its ability to adapt strategies, personalize responses, or optimize processes based on new data provides a significant advantage over static systems.

Before deploying AI agents, organizations must conduct thorough cost-benefit analyses, as agents incur significant costs including development complexity, computational resources for multiple model calls, potential latency, and ongoing maintenance overhead. Additionally, due to their autonomous nature, agents must be rigorously tested in safe, controlled, sandboxed environments to observe behavior, identify failure modes, and refine safety guardrails before interacting with live systems or sensitive data, ensuring reliability, and building organizational trust in their deployment.

Examples of Agents

AI agents are being deployed across various domains. Below examples showcase the breadth of agent applications, each leveraging the core principles of perception, reasoning, and action tailored to specific goals and environments.

- **Customer Service:** Chatbots and voice assistants handle inquiries, schedule appointments, provide support, and personalize interactions. They perceive user queries, reason through natural language processing and goal functions, then respond or make API calls.[7]

- **Autonomous Vehicles:** Self-driving cars, drones, and delivery robots perceive their surroundings through cameras and LiDAR sensors. They reason about navigation and safety using planning and prediction models, then act by steering, braking, or flying to transport goods or people.[8,9]
- **Healthcare:** Diagnostic systems analyze medical images and data. Predictive tools forecast disease outbreaks. Patient monitoring systems track vital signs. Administrative automation handles scheduling and billing tasks.[10]
- **Finance:** Algorithmic trading platforms like those used by Renaissance Technologies analyze market data in real time to make buy and sell decisions.[11] Fraud detection systems identify and block suspicious transactions. Robo-advisors such as Betterment and Wealthfront manage investment portfolios.[12]
- **E-commerce & Entertainment:** Recommendation engines personalize product suggestions on platforms like Amazon or media content on Netflix and Spotify based on user behavior and preferences.[13] Dynamic pricing systems adjust prices based on demand and competitor actions.
- **Supply Chain & Logistics:** Route optimization systems plan delivery paths. Autonomous warehouse robots like those from Kiva Systems manage inventory. Demand forecasting tools predict fluctuations in customer needs.[14]
- **Software Engineering:** Code generation tools such as GitHub Copilot powered by OpenAI Codex assist in writing code. Debugging systems identify errors. Testing frameworks validate complex workflows across multiple files. Deployment automation manages release processes.[15]
- **Smart Homes/IoT:** Home automation systems like Alexa and Google Home manage energy consumption, control lighting

and temperature, monitor security systems, and orchestrate interactions between connected devices based on user preferences and environmental conditions.[16]

- **Scientific Research:** Laboratory automation systems autonomously design and run experiments. Data analysis platforms process vast datasets. Robotic equipment controllers manage complex laboratory instruments.

How Agents Work

What are the underlying mechanisms that enable an agent, whether it is software managing your calendar or a robot exploring Mars, to operate effectively? At its core, an AI agent functions through a continuous cycle involving three fundamental phases: perception, reasoning and action. We dissect this cycle, exploring the key mechanisms and architectures that bring AI agents to life and grant them varying degrees of autonomy. Understanding this cycle is crucial before we delve into the specific types of agents.

Perception

An agent's journey begins with perception—the process of gathering information about its environment. Without perception, an agent is blind and disconnected, unable to make informed decisions. Agents sense their surroundings using various inputs, which can be broadly categorized into physical and virtual sensing mechanisms.

Autonomous agents rely on a wide range of sensory inputs to understand their environment. Physical agents gather data through cameras, microphones, LiDAR, radar, temperature sensors, and GPS, which provide visual, audio, distance, environmental, and location information. In contrast, software agents collect inputs from APIs, databases, network monitors, user interactions, and system logs. These sensory mechanisms—whether detecting physical phenomena or processing digital data streams—provide the foundational perception capabilities that enable agents to understand their operational context and make informed decisions.

A crucial part of perception often involves preprocessing and interpretation. This might include filtering noise from sensor readings, recognizing objects in images through computer vision, understanding human language using natural language processing, or extracting relevant features from large datasets. The outcome of this process is a structured representation of the environment's current state, often referred to as a percept.

Agents often consider the current percept and the entire history of percepts, known as a percept sequence, when making decisions. This historical context becomes particularly important when agents maintain an internal model of the world, allowing them to make more informed and contextually aware decisions based on patterns and changes over time.

Reasoning

After perceiving its environment, the agent moves to reasoning or decision-making. Here, it acts like a brain—processing inputs, combining them with goals, weighing options, and choosing the best action. The complexity of this phase varies enormously depending on the agent's type and architecture, ranging from simple rule-based systems to sophisticated learning algorithms.

Rule-based reasoning represents the simplest form of agent decision-making. These agents operate on predefined if-then rules, also known as condition-action rules. When a certain condition is met in the percept or internal state, the agent executes a specific action. This approach is commonly found in reactive agents that respond directly to environmental stimuli without complex deliberation.[17]

More sophisticated reasoning mechanisms enable agents to handle complex and uncertain environments. Model-based reasoning involves agents maintaining an internal model of the world, which they use to interpret percepts and predict the outcomes of potential actions. Model-based reasoning proves essential for handling situations where the environment isn't fully observable.[18]

Goal-oriented planning takes this further by using planning algorithms, such as search algorithms, to find sequences of actions that lead from the current state to desired goal states. This forward-looking approach allows agents to consider future consequences and develop strategic action plans.[19]

Utility maximization represents an even more advanced reasoning approach where agents evaluate potential actions or states based on a utility function that measures desirability. These agents choose actions expected to yield the highest utility, enabling them to make optimal decisions and handle complex trade-offs between competing objectives.[20]

Learning and flexibility allow many modern agents to incorporate machine learning algorithms into their reasoning processes. These agents learn patterns from data, adapt their internal models or decision rules based on experience and feedback through methods like reinforcement learning, and improve their performance over time. Learning eliminates the need for explicit reprogramming for each possible scenario, making agents more flexible and capable.[21]

The reasoning phase serves as the critical bridge between sensing the world and acting upon it, transforming raw perceptual information into actionable decisions that drive the agent toward its objectives.

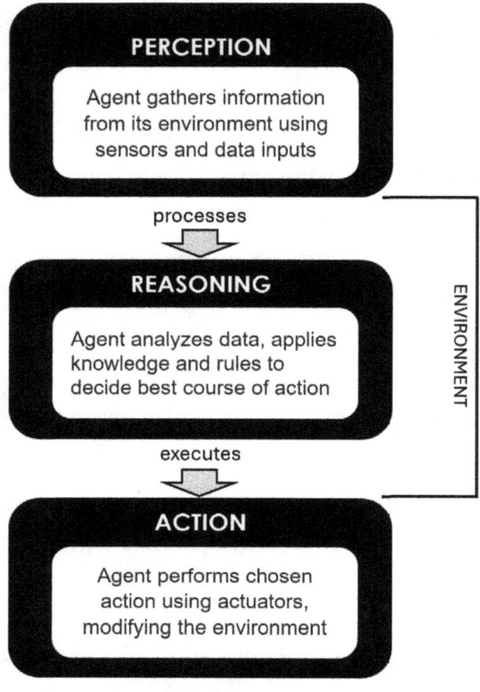

Figure 1: The Agent Loop

Action

The final phase of the basic cycle is action. Having decided what to do, the agent executes its choice, interacting with and potentially changing its environment. This execution is achieved through various types of actuators that serve as the agent's means of influencing the world around it.

Autonomous agents execute actions through actuators designed for their domain. Physical agents use motors, servos, robotic arms, grippers, speakers, displays, and vehicle control systems to manipulate objects and navigate real-world environments. For example, a Mars rover employs its robotic arm to collect soil samples. Software agents, on the other hand, act through messaging platforms, API calls, database updates, user interfaces, transaction systems, and network controls to interact with digital systems and users. A trading algorithm, for instance, executes stock purchases through financial APIs. These actuator mechanisms serve as the essential output channels that transform agent decisions into concrete actions within their respective environments.

The agent's action creates a new state that will be perceived in the next iteration of the cycle. This continuous loop of perception, reasoning, and action enables agents to adapt dynamically to changing conditions and work toward achieving their objectives through sustained interaction with their environment.

Agent Loop Example

A smart thermostat is an example of a simple AI agent.

Perception: The thermostat uses its temperature sensor to gather information about the current room temperature. It perceives that the room temperature is 65°F. This reading is its percept.

Reasoning: The thermostat's internal logic, its brain, processes the percept.

- Its goal is to maintain the room temperature at 70°F.
- Rule-based reasoning guides the system: If current temperature is less than target temperature then turn on heat.
- It considers its internal state. For example, it knows the heating system is currently off and its goal is to maintain 70°F.
- Based on the percept (65°F), being lower than its goal (70°F), it decides to turn on the heat.

Action: The thermostat executes its decision by sending a signal to the heating system, a physical actuator, to turn it on.

This action changes the environment—the room temperature starts to rise. The thermostat then re-enters the perception phase, sensing the new room temperature, and the cycle continues.

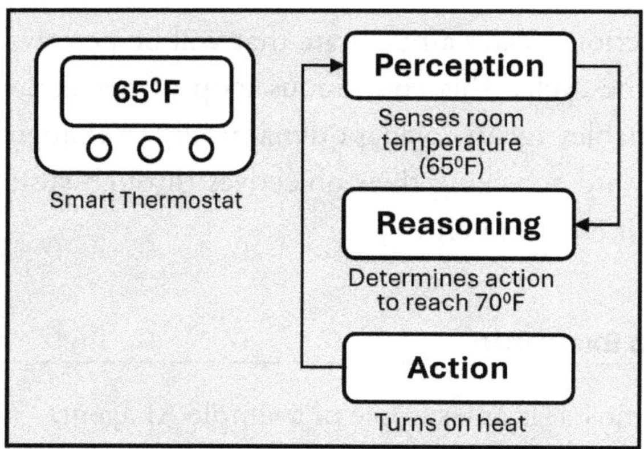

Figure 2: Smart Thermostat Agent Loop

Levels of Agent Autonomy

A defining characteristic of AI agents is autonomy—the ability to operate independently and make decisions without constant human intervention. Autonomy exists on a spectrum from simple rule-based automation to sophisticated independent operation.

Autonomous AI agents operate with high independence. Once given objectives, the agents perceive, reason, learn, and act without requiring human intervention for each step. They adapt to unforeseen circumstances and optimize performance over time. Examples include self-driving cars navigating complex traffic, customer service bots handling diverse inquiries, and predictive analytics tools that continuously update models in healthcare.

Semi-autonomous AI agents perform many tasks independently but rely on human oversight for critical decisions where complete autonomy might be risky or undesirable. Human-in-the-loop requires active participation in decision-making, while human-on-the-loop involves monitoring and intervening only when necessary. Examples include fraud detection systems flagging suspicious transactions for human review, AI-assisted medical diagnostic tools suggesting diagnoses for doctor confirmation, and advanced driver-assistance systems requiring driver supervision.

Achieving the optimal level of autonomy is a critical design consideration that balances the benefits of independent operation with the need for control, safety, and alignment with human intentions. Often, humans provide oversight and set high-level goals while agents manage detailed execution.

With this understanding of how agents work, we are now equipped to explore the rich diversity of agent types and their specific capabilities.

Types of AI Agents

AI practitioners categorize agents to understand their abilities, limitations, and which tasks they are best suited for. Classification helps us choose the proper architecture by selecting an appropriate agent type to tackle a specific problem. The process also provides a framework for comparing the intelligence, autonomy, and complexity of different agents. Additionally, knowing an agent's type helps set realistic expectations of its capabilities, preventing misaligned project goals. Finally, classification guides design and development by informing developers about the necessary components and algorithms required to build effective agents.

We can classify agents based on various criteria, including their internal architecture, the way agents make decisions, their ability to learn, their interaction with the environment, and whether the agents operate alone or with others. While several classification schemes exist in the field, one of the most widely recognized and pedagogically useful approaches categorizes agents based on their complexity and capabilities, ranging from simple reactive machines to sophisticated learning systems. This spectrum-based classification provides a clear progression that helps both newcomers and experienced practitioners understand the evolution and increasing sophistication of agent technologies.[22]

Understanding different types of AI agents is a strategic business decision that impacts your technology investments, implementation timeline, and competitive advantage. Each agent type serves specific business needs and requires different levels of investment and organizational readiness.

Simple Reflex Agents

When you need instant, consistent responses to predictable situations, Simple Reflex Agents are your best option. They excel at automating routine, rule-based decisions that require immediate responses. These systems operate on straightforward if-then logic. When a specific condition occurs, they execute a predetermined action without considering past events or future consequences.[22]

Business Applications: Security alert systems that immediately notify staff when sensors detect unauthorized access, basic customer service chatbots that provide standard responses to common questions, automated inventory alerts when stock levels fall below thresholds, and HVAC systems that adjust temperature based on sensor readings.

Business Decision Criteria:

- ROI Timeline: Up to 3 months
- Implementation Complexity: Low—we can deploy them within weeks
- Data Requirements: Minimal—only current sensor data or inputs are needed
- Risk Level: Low—predictable behavior with clear boundaries
- Investment Range: Low to moderate
- Best for: Well-defined processes, compliance requirements, basic automation

Limitations: These agents cannot handle situations requiring context, memory, or complex decision-making. May create frustrating customer experiences if used inappropriately in complex scenarios.

Model-Based Reflex Agents

When context and history matter for better decision-making you choose Model-Based Reflex Agents. These agents build internal memory of past interactions and environmental states, enabling them to make more informed decisions in situations where current information alone is insufficient. They maintain context across interactions and can infer information that is not immediately observable. [22]

Business Applications: Advanced customer service systems that remember conversation history and customer preferences, navigation systems that track vehicle locations even when GPS signals are temporarily blocked, fraud detection systems that consider account history and transaction patterns, and inventory management systems that account for seasonal trends and supplier reliability.

Business Decision Criteria:

- ROI Timeline: 3-6 months
- Implementation Complexity: Moderate—requires data integration and model design
- Data Requirements: Moderate—historical data and real-time inputs for context building
- Risk Level: Moderate—depends on model accuracy and data quality
- Investment Range: Moderate—for development and data infrastructure
- Best for: Customer relationship management, operational monitoring, pattern recognition

Advantages: This type of agents provide better user experience through context awareness and can handle more complex business scenarios than simple automation.

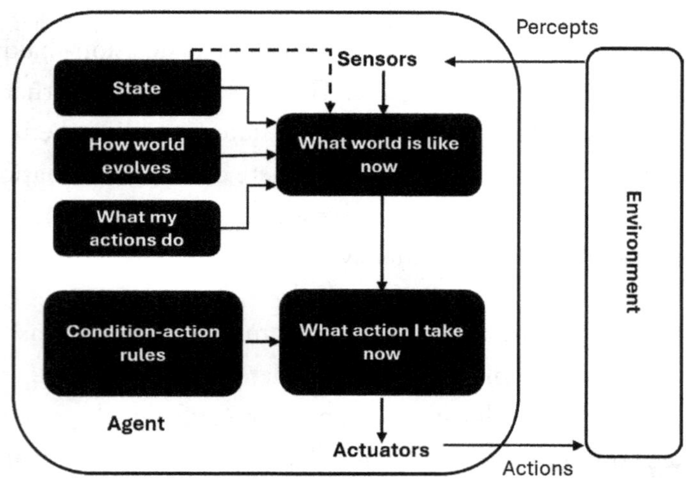

Figure 3: Model-Based Reflex Agents

Goal-Based and Utility-Based Agents

Goal-based agents excel at strategic planning, using multi-step reasoning to achieve specific objectives. Utility-Based Agents, on the other hand, are ideal for optimization across multiple competing priorities. They evaluate trade-offs and maximize overall utility. They go one step beyond goal-based agents, and work by assigning values to possible outcomes and choosing the highest scoring one. Both these agent types represent sophisticated approaches that move beyond reactive responses to proactive planning and optimization.[22]

Goal-Based Applications: Supply chain logistics systems that plan optimal delivery routes across multiple constraints, project management tools that schedule resources to meet deadlines, and financial planning systems that develop strategies to achieve specific targets.

Utility-Based Applications: Investment portfolio management that balances risk and return based on client preferences, dynamic pricing systems that optimize revenue while maintaining competitiveness, and resource allocation systems that balance cost, speed, and quality objectives.

Business Decision Criteria:

- ROI Timeline: 6-18 months
- Implementation Complexity: High—needs domain expertise and sophisticated algorithms
- Data Requirements: Comprehensive data across multiple business dimensions
- Risk Level: Moderate to high—planning decisions can have significant business impact
- Investment Range: High—significant investment in technology and expertise

- Best for: Strategic operations, competitive differentiation, complex optimization problems

Advantages: Both of these agent types can provide substantial advantages through superior planning and optimization, notably in complex business environments with multiple constraints and objectives.

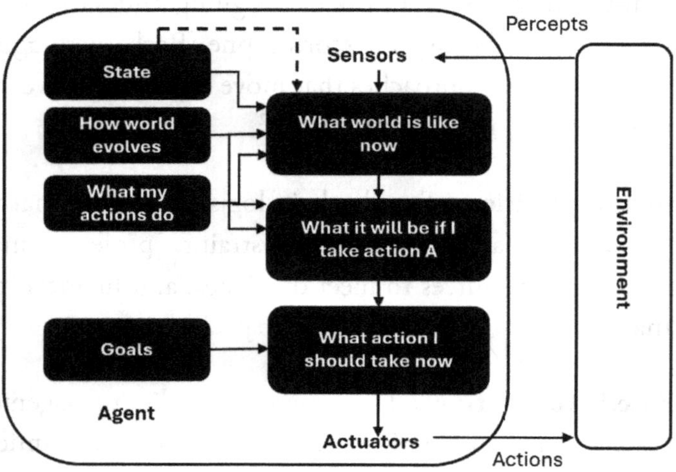

Figure 4: Goal-Based AI Agents

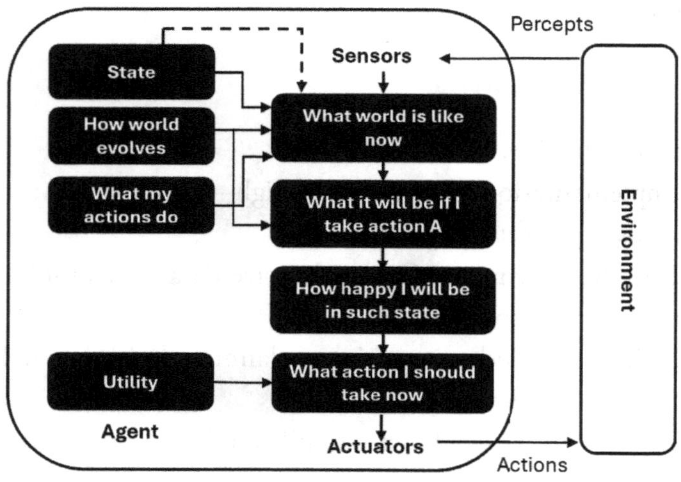

Figure 5: Utility-Based AI Agents

Learning Agents

Learning agents represent the most sophisticated implementation, continuously improving through experience and feedback. They adapt to changing environments and discover new patterns and strategies beyond their original programming. [22]

Business Applications: Recommendation engines that continuously improve suggestions based on user behavior, predictive maintenance systems that learn equipment failure patterns, personalization engines that adapt to individual customer preferences, and trading systems that adjust strategies based on market conditions.

Business Decision Criteria:

- ROI Timeline: 12-24 months or longer
- Implementation Complexity: High--requires ML expertise and continuous monitoring
- Data Requirements: High—large volumes of quality data for training and ongoing learning
- Risk Level: High--requires careful monitoring and governance
- Investment Range: High—substantial investment in technology, talent, and infrastructure
- Best for: Market differentiation, personalization at scale, complex adaptive challenges

Advantages: Highest potential for competitive differentiation and market leadership, particularly in dynamic industries where adaptability is crucial.

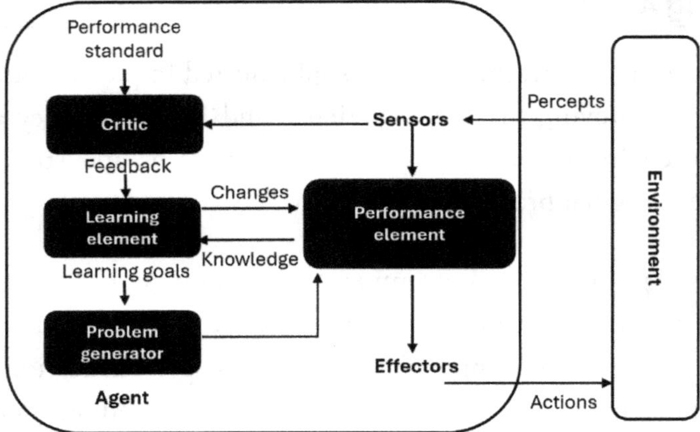

Figure 6: Learning Agents

Choosing the Right Agent Type

Successful AI agent implementation depends on matching the agent type to your specific business needs, organizational capabilities, and strategic objectives. Begin with simpler agent types to build competency, then gradually progress to more sophisticated systems as your data infrastructure and expertise mature.

Many successful AI strategies deploy multiple agent types across different business functions, creating an integrated intelligent system that serves various organizational needs.

Agent Type	ROI Timeline	Complexity	Investment	Best For
Simple Reflex Agents	Up to 3 months	Low	Low - Moderate	Well-defined processes, compliance
Model-Based Agents	3-6 months	Moderate	Moderate	CRM, operational monitoring
Goal/ Utility-Based	6-18 months	High	Significant	Strategic operations, competitive edge
Learning Agents	12-24+ months	Very High	Substantial	Market differentiation, personalization

Table 1: AI Agent Business Decision Framework

2

Agent Architecture: Patterns for Intelligent Systems

"Patterns help you build on the experience of others."
— **Erich Gamma,** Co-author of Design Patterns

My friend, Marcus, had nursed a dream for years—trekking to Everest Base Camp (EBC). Fit and disciplined from his regular gym routine, he possessed the physical foundation but lacked experience with high-altitude trekking. When he asked if AI could help plan this once-in-a-lifetime adventure, I saw an opportunity that would become a perfect case study in intelligent agent design.

What followed went beyond trip planning—it was a comprehensive exploration of how the right architectural patterns can transform AI agents from simple question-answering tools into robust, adaptable companions capable of handling multifaceted challenges. From researching flight options and visa requirements to designing altitude-specific fitness regimens and coordinating tea house reservations, Marcus's EBC journey would reveal how thoughtful agent design patterns create systems that feel nearly human in their support.

Building on our understanding of how agents perceive, reason, and act, we now turn to the architectural blueprints that enable sophisticated AI systems. As software engineers rely on proven design patterns—reusable solutions to recurring problems—AI agent development has evolved its own set of architectural patterns. Architectural patterns are battle-tested approaches for organizing an agent's language model interactions, tools, memory, and reasoning loops.

Think of architectural patterns as the difference between knowing that buildings need foundations and floors versus understanding the precise engineering patterns that make skyscrapers both resilient and functional. These agent design patterns provide detailed blueprints for constructing systems that go beyond merely working to working reliably under complex, real-world conditions.

The most compelling aspect of production-grade agentic systems is their elegant simplicity. Successful agents are typically built from a series of simple, composable patterns—not dependent on monolithic or overly complex frameworks. When developers understand and apply these patterns correctly, the patterns allow them to construct sophisticated behaviors through modular, manageable components.

This chapter illuminates these advanced patterns through Marcus's journey, offering a practical guide to architecting agents that transcend basic intelligence to achieve true robustness and efficiency. We will explore how to decompose complex problems into manageable steps, orchestrate multiple language model calls seamlessly, and ensure our agents can dynamically adapt their strategies to achieve their goals while maintaining the reliability that applications demand.

While this chapter targets professionals with technical backgrounds, readers at all levels will gain valuable insights from the EBC Agent discussions. These insights will help you understand what makes AI agents effective and how to leverage them successfully to solve problems.

Agentic Systems

Before diving into specific patterns, it is crucial to make an important architectural distinction. While this book broadly defines an AI agent as a system that uses an LLM for reasoning, acts via tools, maintains memory, operates with logic, and self-corrects, the way these capabilities are orchestrated can vary substantially. Anthropic categorizes agentic systems into two primary types based on their control structure: workflows and agents.[1]

Workflows are systems where you orchestrate LLMs and tools through predefined code paths. Think of a meticulously planned assembly line where designers explicitly lay out each step. The system might use the LLM at various points to process information or make decisions within a fixed structure, but the developer's code determines the overall sequence of operations.

Consider an e-commerce company's customer support system that processes incoming tickets. This workflow follows a rigid, predictable sequence. The LLM is used at specific steps—extraction and response generation, but the overall process is controlled by predefined business logic. All tickets follow the same path, making them reliable and cost-effective for handling routine support requests.

1. Parse incoming email/chat message
2. LLM extracts key information: order number, issue type, customer sentiment
3. Based on extracted category, route to predefined department
    - Billing issues → Finance team
    - Product defects → Quality team
    - Shipping delays → Logistics team
4. LLM generates appropriate auto-response template
5. Log ticket in CRM system
6. Send confirmation to customer

Figure 7: Workflow Example

Dynamic Agents are systems where the LLM dynamically directs its own processes and tool usage. The LLM itself maintains control over how it accomplishes tasks, deciding which tools to use, in what order, and how to interpret the results to achieve a given goal. This is more like an experienced project manager who adapts their plan on the fly.

Here is an example. The AI agent evaluates a potential acquisition target and provides an investment recommendation. Unlike the workflow, this agent continuously makes decisions about what to investigate next based on what it learns, much like a human analyst would adapt their research strategy as new information emerges.

- Agent searches for TechCorp's recent financial reports
- Discovers concerning revenue decline. Investigates market conditions
- Finds industry-wide downturn, then shifts focus to competitive positioning
- Notices mention of new patents, and researches IP portfolio value
- Discovers regulatory issues, pivots to legal risk assessment
- Synthesizes findings and adapts analysis based on discovered information

Figure 8: Dynamic Agent Example

The choice between workflow and dynamic agent architecture depends heavily on the task at hand, performance, and cost considerations. Use workflows when you need predictability and consistency for well-defined tasks where the steps to achieve a goal are known and repeatable, as workflows offer a reliable and often more straightforward implementation. Opt for dynamic agents when flexibility and model-driven decision-making at scale are essential—especially in situations that are open-ended, involve unpredictable multi-step processes, or demand responsiveness to unforeseen circumstances.

The increased complexity and potentially numerous LLM calls can lead to higher operational costs and slower response times for dynamic agents while offering better performance. Therefore, it is essential to consider if this trade-off is justified for your specific application. For many applications, optimizing single LLM calls with retrieval and in-context examples might be sufficient.

The Augmented LLM

Whether you're building a structured workflow or a dynamic agent, the fundamental building block is an LLM that has been enhanced with various augmentations. Unlike a plain LLM, this augmented LLM possesses essential capabilities that enable practical functionality.

In the Everest Base Camp Agent example, static knowledge about EBC wasn't sufficient—the agent needed to access current weather forecasts, check real-time flight prices, retrieve visa requirements from Nepal's government website, and utilize training tools to customize a fitness plan based on Marcus's specific age and experience. These actions demanded capabilities far beyond the static knowledge embedded in the LLM's pretraining.

The key augmentations that enable LLM functionality are:

- **Retrieval:** The ability to access and fetch information from external knowledge bases such as a company's internal documents or a public search engine to inform its responses or decisions. This is a core aspect of Retrieval Augmented Generation (RAG).
- **Tools:** The capacity to use external tools or APIs to perform actions or gather information beyond its inherent knowledge. This capacity ranges from sending emails and querying databases to running code and accessing real-time data like stock prices.
- **Memory:** Mechanisms to maintain context over multiple steps or learn from past interactions, enabling more coherent and adaptive behavior.

Modern LLMs, like those from OpenAI, Google, and Anthropic, can actively utilize these augmentations. They can generate their own search queries for retrieval, select appropriate tools from a provided set, and even determine what information is crucial to retain in memory for future steps.

When implementing these augmentations, you should tailor them to your specific use case and ensure they provide an easy, well-documented interface for the LLM. Protocols like the Model Context Protocol (MCP), for instance, aim to simplify the integration of third-party tools by allowing developers to connect to an ecosystem of tools via client implementation.[2]

Throughout the patterns discussed below, we will assume that each LLM call has access to these augmented capabilities.

Workflow Patterns

Many sophisticated agentic behaviors can be constructed by composing simpler, well-defined workflow patterns. These patterns help structure the interaction between LLMs, tools, and data, often leading to more reliable and understandable systems. Here are common patterns from successful production systems, based largely on insights from Anthropic.[1]

Prompt Chaining

Prompt chaining decomposes a task into a sequence of distinct steps, where each LLM call processes the output of the previous one. It is like a relay race where the baton (the output) is passed from one runner (LLM call) to the next.

An important feature of prompt chaining is the ability to insert programmatic checks or gates between steps. These gates can validate intermediate outputs, ensure the process is on track, or even route the workflow differently based on the output of a previous step. For example, after an LLM generates an outline, a gate could check if the outline meets specific criteria before passing it to the next LLM call for document generation.

Conceptual Flow: Input -> LLM Call 1 (Subtask 1) -> Output 1 | Gate 1 -> LLM Call 2 (Subtask 2, using Output 1) -> Output 2 | Gate 2... -> Final Output

When to Use: Use this pattern when a task can be cleanly broken down into fixed subtasks. The primary goal is often to trade off some latency for higher accuracy by making each individual LLM call simpler and more focused.

Examples: One example involves generating marketing copy in one language, then passing that copy to another LLM call for translation into a different language, allowing each step to focus on its specific expertise—content creation and translation, respectively. Another powerful application is writing an outline for a document, programmatically checking if the outline meets certain criteria through a validation gate, and then having a subsequent LLM call write the full document based on the validated outline, ensuring both structural integrity and content quality.

In our EBC Agent case, Prompt Chaining looks like this:

Visa Research → Timeline → Flight Booking

For illustration purposes, we only show the Visa Research step. Based on the Visa processing time, Timeline will be adjusted.

- Input: "Marcus needs to trek to EBC in April 2025"
- Output: "Nepal tourist visa required. Embassy visa recommended (10 business days processing)"
- Gate: Processing time validation → Adjust departure timeline

Trade-offs: Prompt chaining improves accuracy by breaking complex tasks into focused subtasks, enabling better control and intermediate validation to catch issues early. However, this approach increases latency through sequential LLM calls and raises costs due to higher token usage from multiple API requests.

Routing

Not all customer inquiries are the same. Some are simple FAQs, others are complex technical issues, and some might be urgent complaints. Trying to handle all these with a single, generic prompt can lead to suboptimal responses. Routing offers a solution.

Routing involves an initial LLM call or a traditional classifier that classifies an input and then directs it to a more specialized downstream task, prompt, or even a different LLM model.

> Conceptual Flow: Input -> LLM Call (Classifier) -> [Route A -> Specialized LLM/Prompt A OR Route B -> Specialized LLM/Prompt B]

This workflow lets you create specialized prompts and tools for different types of tasks. When you optimize a prompt for one specific input, it often performs worse on other types of inputs. Routing solves this problem by directing each task to its purpose-built prompt.

When to Use: Routing excels at complex tasks containing distinct input categories that require separate handling, where the system can accurately classify inputs initially.

Examples: A powerful application involves optimizing for cost and speed by routing easy or common questions to smaller, faster models while directing complex or unusual questions to more capable models that may be slower or more expensive, ensuring that computational resources are allocated appropriately based on the complexity of each request.

For the EBC Agent, Routing looks like this:

> Query Analysis → Handler Classification → Specialized Processing

Based on the query type, the Classifier routes the request to the appropriate specialized handler.

- Input: "What sleeping bag should I bring for EBC?"
- Output: "GEAR query detected → Route to Equipment Specialist"

Trade-offs: Routing improves performance by using specialized prompts for different input types and enables cost optimization by matching models to task complexity. However, it adds latency and expense through the classification step, and the entire system's effectiveness depends on accurate initial routing—misclassified requests can degrade user experience and require additional correction.

Parallelization

Parallelization involves one or more LLMs working on different aspects of a task simultaneously, with their outputs potentially being aggregated programmatically.

> **Sectioning:** Breaking a task into independent subtasks that can be processed in parallel and then combined. For example, when writing a long report, different sections—Introduction, Methodology, Findings, Conclusion--could be assigned to parallel LLM calls.
>
> > Conceptual Flow: Input -> [LLM Call A (Part 1) || LLM Call B (Part 2) || LLM Call C (Part 3)] -> Aggregate Outputs -> Final Output

Voting: This method involves running the same task multiple times, potentially with slightly different prompts or parameters, in order to obtain diverse outputs. These outputs can subsequently be compared, validated, or aggregated. Examples include taking the majority vote for a classification task or selecting the best response based on evaluation criteria.

> Conceptual Flow: Input -> [LLM Call A (Attempt 1) || LLM Call B (Attempt 2) || LLM Call C (Attempt 3)] -> Evaluate & Select/Aggregate -> Final Output

When to Use: Parallelization is effective when subtasks can be run independently to save time through Sectioning or when multiple perspectives or attempts are needed for higher confidence, robustness, or quality through Voting. Anthropic notes that for complex tasks with multiple considerations, LLMs generally perform better when each consideration is handled by a separate LLM call, allowing focused attention.[1]

Examples: Parallelization can be effectively implemented through both Sectioning and Voting approaches across various practical scenarios. In Sectioning applications, implementing safety guardrails demonstrates clear benefits. One LLM instance processes a user query while another simultaneously screens it for inappropriate content, often performing better than a single LLM attempting both tasks. This approach also works well for automating LLM performance evaluations, where different LLM calls evaluate distinct aspects such as coherence, factuality, and tone of a model's response to a given prompt.

Voting approaches prove valuable in scenarios like reviewing code for vulnerabilities, where several different prompts or LLM calls examine the code and an issue is flagged if any or a majority find a problem, and evaluating whether content is inappropriate through multiple prompts that assess different aspects or use different vote thresholds to balance false positives and false negatives.

Coming to our EBC Agent, it needs to research flight prices and weather research which can run in parallel. In this case, Parallelization looks like this:

> Task Analysis → Concurrent Processing → Result Integration

For illustration purposes, we show the concurrent execution step only.

- Input: "Plan EBC trek for Marcus in October 2025"
- Output: "PARALLEL BRANCH A: Flight pricing || PARALLEL BRANCH B: Weather analysis"
- Aggregate: Synchronization → Wait for both branches to complete before integration

Trade-offs: Parallelization can improve speed through concurrent execution and enhance quality by leveraging multiple perspectives or attempts, notably in voting scenarios where diverse approaches catch issues a single attempt might miss. However, it increases computational costs by multiplying LLM calls and adds system complexity through the need for output aggregation logic.

Orchestrator-Workers

For tasks where the sub-steps aren't known in advance, a dynamic, hierarchical approach is needed. The Orchestrator-Workers pattern enables this. In this workflow, a central orchestrator LLM dynamically breaks down a complex task into smaller subtasks. It then delegates these subtasks to worker LLMs which could be instances of the same model or different, specialized models. Finally, the orchestrator synthesizes the results from the worker LLMs to produce the final output.

> Conceptual Flow: Input -> Orchestrator LLM (Plans & Decomposes) -> [Worker LLM 1 (Subtask A) -> Result A || Worker LLM 2 (Subtask B) -> Result B] -> Orchestrator LLM (Synthesizes Results) -> Final Output

In the Orchestrator-Workers pattern, the orchestrator LLM determines subtasks on-the-fly based on the specific input and evolving problem state, in contrast to Sectioning which breaks tasks in a predefined way.

When to Use: This pattern is well-suited for complex, dynamic tasks where you can't predict the exact sequence or number of subtasks required.

Examples: Coding agents exemplify this approach when making complex changes across multiple files based on a high-level task description, where the orchestrator identifies which files need editing and what kind of changes are required, then assigns those specific editing tasks to specialized workers who can focus on particular aspects of the codebase. Similarly, research or search tasks that involve gathering, analyzing, and synthesizing information from multiple diverse sources benefit from this pattern. These tasks often require answering complex queries where the orchestrator can dispatch workers to search different databases or websites based on the evolving needs of the investigation. The orchestrator then synthesizes their findings.

Here is how the Orchestrator-Workers pattern looks like for the EBC Agent for altitude training design:

> Task Assessment → Worker Delegation → Result Synthesis

The delegation step looks like this:

- Input: "Marcus needs altitude training for EBC (5,364m), timeline: 12 weeks"
- Output: "DELEGATE → Fitness Specialist Worker with parameters: 45-year-old, current fitness level, altitude target" || "DELEGATE → Local Trek Specialist Worker with parameters: resides in Akron, OH, allergic to pollen"

Trade-offs: The Orchestrator-Workers pattern excels at handling complex, unpredictable problems by dynamically breaking them into manageable pieces without predetermined workflows, making it ideal for open-ended tasks where the solution path emerges during execution. However, it is challenging to implement and debug, incurs high latency and costs from multiple LLM calls, and depends heavily on the orchestrator's ability to effectively decompose and manage tasks.

Evaluator-Optimizer

The Evaluator-Optimizer pattern proves valuable in scenarios requiring iterative refinement to achieve high-quality outputs. This pattern embraces iterative approach by creating a loop of generation and critique. In this workflow, the Optimizer generates an initial response or artifact such as a summary or a piece of code. The Evaluator then critiques the artifact against defined criteria. This feedback is then used by the Optimizer in a subsequent call to refine its output. This loop can continue for a set number of iterations or until the output meets a certain quality threshold.

> Conceptual Flow: Input -> Optimizer LLM (Generates v1) -> Evaluator LLM (Critiques v1) -> Feedback -> Optimizer LLM (Generates v2 using Feedback) -> ... -> Final Polished Output

This is analogous to the iterative writing process a human might go through, producing drafts, getting reviews, and revising until a polished document is achieved.

When to Use: This pattern is effective when clear evaluation criteria exist and when iterative refinement demonstrably adds value. You will know it is a good fit if two conditions are met. First, LLM responses get measurably better when a human gives feedback on them. Second, an LLM can provide useful, constructive feedback on its own.

Examples: Literary translation exemplifies this approach, where an initial translation from the optimizer might miss cultural nuances or subtle meanings that an evaluator LLM acting as a critic can identify, leading to subsequent refinements that capture the original text's true essence. Complex search or research tasks also benefit from this pattern through multiple rounds of searching and analysis, where the evaluator LLM determines whether current information is comprehensive enough or if further searches by the optimizer/searcher LLM are warranted to fully address the query. Code generation represents another strong use case.

For Marcus's EBC Agent, an Optimizer could generate an initial 12-week fitness plan. An Evaluator agent could then critique it against Marcus's specific biometrics and high-altitude safety protocols, providing feedback like, 'The cardio progression is too aggressive for a novice trekker. Recommend adding two more weeks of Zone 2 training to build aerobic base.' The Optimizer would then refine the plan based on this feedback.

Trade-offs: The Evaluator-Optimizer pattern delivers higher quality, more precisely aligned outputs by mimicking human iterative refinement processes, achieving polish that single-pass generation often cannot match for complex tasks. However, the Evaluator-Optimizer pattern comes at significant cost through high latency and expense from multiple LLM calls, challenges in defining reliable evaluation criteria, and risks of endless refinement loops that consume resources without meaningful improvement.

Workflow Pattern	When to Use	Key Benefit
Prompt Chaining	Tasks with clear sequential steps	Higher accuracy through focused subtasks
Routing	Multiple input types needing specialized handling	Optimized performance via specialization
Parallelization	Independent subtasks or need for multiple perspectives	Speed (Sectioning) or quality (Voting)
Orchestrator-Workers	Complex, unpredictable tasks	Maximum flexibility for dynamic problems
Evaluator-Optimizer	Tasks requiring iterative refinement	Highest quality through feedback loops

Table 2: Workflow Pattern Selection Guide

Dynamic Agents

Dynamic agents bring capabilities a step further than the structured workflows described in the previous section by placing the LLM in more direct control of its own processes and tool usage in a loop, guided by environmental feedback. These systems are emerging as LLMs mature in capabilities like complex input understanding, reasoning, planning, reliable tool use, and error recovery.[1,3]

An autonomous agent typically starts with a command or an interactive discussion with a user to clarify the task. Once the goal is clear, the agent plans and operates with a degree of independence, deciding which tools to call, interpreting their outputs, and adjusting its plan as needed. It is crucial for these agents to get facts from the environment at each step, such as results from a tool call or output from code execution, to assess progress and self-correct. They might pause for human feedback at checkpoints or when encountering blockers.

Dynamic agents are best for open-ended problems where predicting the required number of steps or a fixed path is difficult. The LLM may operate for many turns, requiring a level of trust in its decision-making capabilities. Their autonomy makes them suitable for scaling tasks in trusted environments. Despite their sophisticated behavior, the core implementation is often an LLM using tools based on environmental feedback within a loop. This makes the clear design and documentation of tools absolutely critical for the Agent-Computer Interface, a topic we will explore in depth later in this chapter.

Examples: A coding agent designed to resolve software engineering tasks described in natural language exemplifies the Dynamic Agent approach, as it might involve reading multiple files, writing code, running tests, and debugging based on test results, all orchestrated by the agent's own reasoning capabilities that adapt to the specific requirements and challenges encountered during execution. Similarly, a computer use agent that can operate a computer through APIs simulating mouse clicks, keyboard input, file operations, and web browsing represents another powerful application. The agent enables the system to accomplish complex tasks by autonomously navigating digital environments and making decisions.

Trade-offs: Dynamic agent workflows provide maximum flexibility and true autonomy for complex, unpredictable tasks, dynamically responding to changing conditions without predetermined sequences or constant human guidance. However, this autonomy introduces significant risks including higher costs from extensive LLM calls, compounding errors from poor early decisions, and potential runaway processes. These challenges require extensive sandboxed testing and robust guardrails like iteration limits to maintain control and prevent excessive resource consumption.

Combining Patterns

These design patterns become more powerful when combined and customized for specific use cases. A routed workflow might trigger a sub-workflow using prompt chaining, which includes an evaluator-optimizer loop for quality refinement. An orchestrator could delegate tasks to workers that employ their own prompt chains or parallelization strategies. For instance, a customer service AI system might use routing to classify incoming requests, then delegate complex technical issues to a specialized worker that employs prompt chaining to gather system information, followed by an evaluator-optimizer loop to refine and validate the proposed solution before presenting it to the customer.

As we emphasized before, start with the simplest solution and add complexity only when it demonstrably improves outcomes. Each additional layer and LLM call increases latency and cost, making rigorous performance measurement and continuous iteration essential for finding the optimal balance.

Here are the key trade-offs to evaluate.

Performance vs. Cost: Complex patterns with multiple LLM calls typically deliver better results but invariably increase execution time and expenses. The critical consideration is whether the performance gain justifies the additional cost.

Flexibility vs. Predictability: Dynamic agents provide maximum adaptability but sacrifice the predictability of structured workflows. Your choice depends on tolerance for variability versus the need for strict process adherence.

Development Complexity: While frameworks simplify pattern implementation, building and debugging complex, nested agentic systems presents significant challenges. Sometimes a few direct lines of LLM API code implementing a simple pattern prove more transparent and maintainable. When using frameworks, understanding the underlying code and prompts is crucial to avoid incorrect assumptions about system behavior.

Advanced State Management

AI agents require sophisticated state management to maintain memory of pertinent information over time, including current situational understanding, past actions, goal progress, and learned knowledge. The chosen strategy depends on the agent's complexity, required context duration, and information type. Poor state management causes agents to appear forgetful, repeat actions, or fail at tasks requiring coherent long-term planning.[4]

Effective memory is what enables agents to handle complex multi-step workflows, learn over time, and recover from errors. Without it, agents cannot remember recent actions, learned information, or previous user interactions, severely limiting their capabilities.

Memory Requirements Across Patterns: Different agent patterns have distinct memory needs. For example, Orchestrator-Worker systems require the orchestrator to track delegated subtasks, assigned workers, and returned results. Evaluator-Optimizer loops need the optimizer to remember previous artifact versions and feedback for improvement. Dynamic agents must maintain their plans and action results to inform subsequent decisions.

Short-term Memory: Working memory focuses on effective context window management through conversational history techniques that pass relevant prior interactions to the LLM, and scratchpad approaches that enable step-by-step thinking by recording intermediate thoughts and feeding them back into subsequent prompts—closely related to Chain-of-Thought prompting, which we discuss in the next section.[5]

Long-term Memory Systems: Persistent memory systems retain information across sessions and access knowledge beyond working memory limits. Retrieval Augmented Generation (RAG) connects agents to vector databases containing document or experience embeddings, enabling relevant information retrieval to augment current context. Knowledge graphs store structured entity relationships for efficient querying, while traditional databases and file systems provide reliable persistence for cross-session information maintenance.[5]

Implementation Frameworks: LangChain reduces implementation complexity by offering pre-built components and abstractions for different memory types. These frameworks handle underlying complexity, allowing developers to focus on higher-level agent logic while leveraging robust, tested memory management systems.[6] The Akka platform provides a powerful actor-based model well-suited for distributed agent systems, where each actor maintains its own state and communicates through message passing, offering natural isolation, fault tolerance, and scalability for complex multi-agent architectures that require persistent state management across distributed environments.[7]

EBC Agent's State Management: Here is how Marcus's EBC Agent leverages state management across multiple dimensions. The agent leveraged short-term memory through conversational history to track Marcus's evolving preferences—remembering his spring or fall availability constraint while dynamically pivoting from April to October based on crowd analysis. Scratchpad approaches maintained complex reasoning chains when the agent discovered visa processing timelines, requiring it to adjust the entire planning sequence while preserving budget constraints and experience level considerations.

For long-term persistence, the agent employed RAG systems to access real-time flight prices and weather forecasts, while maintaining structured knowledge about trekking seasons, altitude requirements, and Marcus's fitness profile across multiple planning sessions. This multi-layered memory architecture enabled the agent to handle the trip's interconnected complexities—coordinating flights, accommodations, training regimens, and visa requirements—while adapting dynamically to new constraints without losing sight of Marcus's budget and novice trekking experience.

Advanced Reasoning and Orchestration

At the core of any advanced AI agent is a dynamic Orchestration Layer, a concept that elevates it from a simple tool to a thinking, acting, and adapting partner. This layer functions as a continuous loop that plans, acts and adjusts. It is the engine room where the agent's thought process comes alive. In this section, we will delve into the key LLM-centric techniques that power this orchestration.

While a single call to a powerful LLM can summarize information or answer questions, building an AI agent that can solve multi-step problems, interact with its environment via tools, and adapt to new information requires more than isolated LLM invocations.

The core challenge is to structure and guide the LLM's reasoning capabilities through a series of steps, allowing it to:

- Decompose complex goals into manageable sub-goals.
- Gather necessary information, often by using external tools.
- Maintain context and track progress across multiple steps.
- Evaluate different options or strategies.
- Learn from its actions and adapt its approach.
- Self-correct when things go wrong.

The mechanisms we explore here provide structured ways to harness LLM's reasoning power for these purposes.

Chain of Thought

One of the foundational techniques for enhancing an LLM's reasoning ability is Chain-of-Thought (CoT) prompting. This approach encourages the LLM to break down a complex problem into a series of intermediate reasoning steps before arriving at a final answer. Instead of directly outputting the solution, the model first thinks step-by-step articulating its rationale. For an agent, CoT can be used internally by the LLM when it needs to reason about a complex sub-problem, plan a short sequence of actions, or decompose a goal. The thought generated by CoT can then inform the agent's next decision or action.[8]

CoT prompting improves LLM reasoning by encouraging the model to show its work step-by-step. This is achieved either through few-shot examples that demonstrate the problem-solving process, or through zero-shot prompting with phrases like "Let's think step by step."

For example, consider this problem: "Roger has 5 tennis balls. He buys 2 more cans of tennis balls. Each can has 3 balls. How many tennis balls does he have now?" A CoT approach would produce: "Roger started with 5 balls. He bought 2 cans, and each can has 3 balls, so that's 2 × 3 = 6 more balls. In total, he has 5 + 6 = 11 tennis balls."""

Specific to our EBC Agent, CoT looks like this. Marcus needs gear for high altitude. Let me think step by step: He's going to 5,364m elevation → oxygen is 50% of sea level → he will need insulation layers → temperatures can drop to -20°C → he will need a sleeping bag rated to -25°C minimum."

The benefits of using CoT are twofold. First, for tasks requiring arithmetic, common sense, or symbolic reasoning, research shows that CoT greatly improves the accuracy of LLM outputs. Second, the articulated reasoning steps provide transparency, offering insight into how the LLM arrived at its solution and making it easier to debug errors or understand the model's logic.

Tree of Thoughts

While Chain-of-Thought follows one reasoning path, Tree of Thoughts (ToT) explores multiple paths simultaneously. ToT structures problem-solving as a tree search, where each node represents a partial solution and branches show different possible next steps. This method allows the LLM to consider various approaches before choosing the best one.[9]

The process begins with thought generation, where the LLM creates multiple distinct thoughts or continuations from its current state. Then, either the LLM itself or an external heuristic evaluates the promise of these generated thoughts. This evaluation might involve checking for coherence, progress toward the goal, or the likelihood of success. A search algorithm, such as a breadth-first or depth-first search, navigates this tree of thoughts, deciding which branches to explore further based on the evaluations.

For example, a game-playing agent could use ToT to consider several possible next moves, evaluate the board state resulting from each one, and then explore deeper sequences from the most promising options. Similarly, a creative writing agent could generate multiple plot points or dialogue options, evaluate their consistency or impact, and then develop the story along the most compelling path.

Marcus's EBC Agent demonstrates Tree of Thoughts process when exploring multiple route options for an October departure. Rather than following a single planning path, the agent generates multiple routing branches—the traditional Lukla flight approach, the weather-independent Jiri Road route, and the less crowded Phaplu airstrip alternative. Each branch is evaluated against criteria like flight reliability, crowd levels, and acclimatization schedules. For weather windows, the agent explores three October timeframes simultaneously. These include early October with post-monsoon clarity and crisp temperatures, mid-October offering optimal conditions despite peak season crowds, and late October featuring clear skies, colder nights, and fewer trekkers. After systematic evaluation of all branches, the agent selects the optimal combination: Phaplu starting point with early-October timing.

This method offers several advantages. It is effective for problems that require exploration, strategic lookahead, or where the initial steps might be uncertain and lead to dead ends. It allows for a more systematic exploration of the solution space compared to a single chain of thought. Furthermore, the evaluation step enables the system to prune unpromising branches and backtrack if a particular line of reasoning proves fruitless, providing a form of self-correction.

For AI agents, ToT provides a more robust mechanism for performing complex planning, making strategic decisions in the face of uncertainty, or solving problems where a single line of reasoning may not be enough.

Tree of Thoughts empowers agents to be more deliberate and strategic, moving beyond linear thought processes to explore a richer solution space and make more informed choices.

ReAct

For agents that need to interact with an external environment or use tools to gather information, reasoning alone is not enough. They must also act. The Reason and Act (ReAct) framework provides a powerful paradigm for this by interleaving reasoning and acting. ReAct enables an LLM to generate both reasoning traces, which are thoughts about what to do, and actions, which are specific tool calls. This process creates a dynamic loop where the agent thinks, acts, learns from the action's outcome, and then thinks again to plan the next step.

The ReAct loop has three components. First is the Thought, where the LLM analyzes the current task and its internal state, including past observations, to generate a thought about its strategy, what information is missing, or what it should do next. Based on this thought, the LLM then moves to the second step, Action, which is often a call to an external tool like a web search API, a database query, or a calculator. This action specifies the tool's name and the necessary parameters. Finally, the agent executes the action and receives an Observation, which is the result or output from the tool. This Observation provides new information or feedback from the environment and is fed back into the LLM in the next iteration, influencing its subsequent thought and action. This cycle continues until the agent believes it has accomplished its goal.

For example, for Marcus's EBC planning, the ReAct loop might proceed as follows when researching accommodations. Thought: "I need to check tea house availability along the EBC route for October." Action: Search tool query for "EBC tea house availability October 2024

Namche Tengboche." Observation: "Namche tea houses are typically full by October 15th. Tengboche has limited options. Advance booking is recommended through local operators." Thought: "Limited availability means I should recommend booking immediately and provide backup accommodation strategies." This ReAct cycle enables the agent to gather real-time information and adapt its accommodation recommendations based on actual availability, not generic advice.

This framework offers several key benefits. It allows agents to interact dynamically with external environments and tools, gathering information as needed. This information enables them to adapt their plans based on the outcomes of their actions. The process also grounds the agent's reasoning in tangible information obtained through tool use, which can reduce the likelihood of errors. Furthermore, the explicit thought traces make the agent's decision-making process more transparent and interpretable.

ReAct is a fundamental orchestration mechanism for many modern, tool-using AI agents. It provides the core loop for agents that need to explore, gather information, and interact with external systems to solve problems.

Reflection

Intelligent agents must not only execute tasks but also learn from their experiences and improve over time. Reflection techniques enable agents to review their past actions, identify errors or inefficiencies, and use these insights to enhance future performance or correct their current course of action. The ability to learn from experience supports the idea of an agent that self-corrects when things go wrong. Reflection involves an agent, often guided by an LLM, critically examining its own performance trajectory—a sequence of thoughts, actions, and observations—to derive lessons or generate feedback for improvement.[10]

The Reflection process begins with storing the agent's interactions, including its reasoning steps, tool calls, and the results from those tools. After completing a task or encountering a failure, the LLM is prompted to reflect on this recorded trajectory. This prompt might ask it to identify what went well, what went wrong, why an error occurred, or how it could have performed the task more efficiently. The LLM then generates reflective insights, which could be high-level feedback, specific corrections to its plan, or even suggestions for modifying its internal prompts or strategies. These insights are then used to guide the agent's future behavior, for instance, by being added to its prompt as context for subsequent tasks or used to update a knowledge base of learned lessons.

For example, a coding agent tasked with writing a Python function might generate code and then run automated tests, only to see them fail. The agent can then reflect on the error messages and its generated code, identify a bug in its logic, and then attempt to regenerate the code with the necessary correction. Similarly, a customer service agent, after a series of interactions, might reflect on which responses led to higher customer satisfaction scores and adjust its communication style accordingly for future conversations.

In Marcus's case, the EBC Agent demonstrates reflection when it initially recommended a standard -10°C sleeping bag based on typical early October temperatures, only to discover through Marcus's feedback about his cold sensitivity and budget constraints that this recommendation was suboptimal. Upon reflection, the agent analyzed its trajectory: it had focused solely on temperature ratings without considering individual comfort preferences or price limitations. The agent generated reflective insights: "I should incorporate user-specific comfort factors and budget constraints earlier in gear selection, beyond technical specifications." This reflection led the agent to adjust its approach for future recommendations, now asking about cold tolerance and budget parameters before suggesting specific gear, demonstrating continuous improvement in its advisory capabilities.

Reflection offers several significant benefits. First, it enables agents to learn from both successes and failures without the need for explicit retraining of the underlying LLM. Secondly, it helps agents become more robust and resilient by learning to avoid past mistakes. Over time, agents can also learn to optimize their strategies and tool usage, leading to enhanced efficiency. Crucially, Reflection provides a direct mechanism for agents to identify and potentially recover from errors mid-task. Ultimately, reflection is crucial for achieving more autonomous, adaptive, and continuously improving agentic systems.

Feature	Purpose	When to Use
Chain-of-Thought (CoT)	Step-by-step reasoning	Internal planning and problem decomposition
Tree of Thoughts (ToT)	Explore multiple paths	Complex planning under uncertainty
ReAct	Reasoning + Action loop	Tool-using agents in dynamic environments
Reflection	Learn from experience	Autonomous agents needing continuous improvement

Table 3: Reasoning Techniques

Integrated Reasoning for Agent Orchestration

CoT, ToT, ReAct, and Reflection techniques are not mutually exclusive and often work together to enable the complex reasoning loops characteristic of sophisticated AI agent behavior. For instance, CoT can structure the thought phase within a ReAct loop, aiding in tool selection or observation interpretation. ToT can be applied when an agent needs to explore multiple thought branches before deciding on an action or plan at critical decision points. Furthermore, Reflection allows for the analysis of an entire ReAct loop's trajectory, or sequences of CoT/ToT steps, to refine the agent's overall strategy or future prompting.

These mechanisms form an agent's Orchestration Layer that manages four key functions: breaking down complex goals into strategies (CoT and ToT), interacting with tools and environments (ReAct), processing feedback and new information (ReAct and Reflection), and adapting plans based on results (ReAct, Reflection, and ToT). Together, they help the agent maintain its internal state—tracking what it knows, what it aims to achieve, and what it has tried—while coordinating LLM interactions to produce coherent, purposeful behavior across multiple reasoning steps.[1]

However, these advanced reasoning mechanisms come with significant trade-offs: higher costs and latency from multiple LLM calls, complex prompt engineering requirements, increased risk of compounding errors, and challenging debugging processes. While they can improve task performance, the added complexity and expense should only be justified when simpler methods are insufficient.[11]

With a grasp of how agents can reason and plan, our focus now shifts to the tools they wield.

Agent-Computer Interface

Advanced AI agents rely on tools to perform complex tasks like fetching information, making calculations, and executing actions. However, effective tool use depends critically on how these tools are designed and presented to the agent's underlying LLMs. This is the domain of the Agent-Computer Interface (ACI), which focuses on making tools intuitive for AI agents, much like Human-Computer Interface (HCI) does for human users. As experts at Anthropic note, tool definitions and specifications should be given as much prompt engineering attention as your overall prompts, and we should plan to invest as much effort in creating good ACI as in HCI.[1]

Critical Role of ACI

Imagine handing a skilled craftsperson a new, complex tool with a poorly written or misleading manual. Their skill might be immense, but their ability to use that specific tool effectively would be severely hampered. Similarly, an LLM, no matter how powerful its reasoning capabilities, relies entirely on the ACI to understand and interact with its available tools.

The Agent-Computer Interface dictates how the agent understands and interacts with its available tools. The definition and description of a tool are the primary ways an LLM learns the tool's purpose and when it might be relevant to the task at hand. Furthermore, the ACI specifies how to call the tool correctly by outlining the necessary parameters, their required format, and the kind of output to expect, which minimizes the chances of incorrect tool calls. Consequently, a well-designed ACI directly contributes to the agent's reliability and its ability to achieve its goals. A poorly designed interface, however, with vague descriptions or confusing parameters, can lead the agent to misuse tools, fail at tasks, or produce incorrect results, a risk that is problematic in autonomous agents where errors can compound.

Even the most sophisticated LLMs are not mind-readers. They operate based on the information provided to them. Therefore, a thoughtfully crafted ACI is more than good practice. It is a fundamental requirement for building effective, robust, and intelligent agents that can reliably leverage external functionalities.

Best Practices for Tool Definition

Defining a tool for use by an LLM is similar to writing a clear API specification or an ideal function docstring—but with the LLM as the end user. As Anthropic suggests, think of this as writing a great docstring for a junior developer on your team.[12] The objective is to minimize ambiguity and make it as easy as possible for the LLM to understand the tool's purpose, usage context, and how to invoke it correctly. To achieve this, follow these best practices.

Comprehensive Descriptions: The tool description acts as the LLM's first and most critical clue. It must be unambiguous, specific, and concise. If you have multiple tools with overlapping functionality, ensure each has a distinct and well-differentiated description. For instance, instead of generically labeling two tools as "file editor," use specific names like file_editor_append_line and file_editor_replace_text to reflect their unique purposes. This helps the model select the correct tool.

Well-Defined Parameters: Each parameter should have an intuitive name and a detailed explanation. Clearly indicate input formats (e.g., "YYYY-MM-DD" for dates), expected types, and whether a parameter is required or optional. Specify constraints when necessary—for example, if a file path is needed, define whether it should be absolute or relative. Anthropic's research with their SWE-bench agent found absolute paths to be more reliable.[1]

Illustrative Examples: Provide concrete examples of tool invocation, including sample inputs and expected outputs. Those examples demonstrate the tool's behavior and give the LLM a working model to emulate.

Edge Cases and Boundaries: Explicitly state what the tool does not do. Define edge cases, limitations, and input constraints to prevent misuse. For example, if a tool only handles structured data, say so. When similar tools exist, please make sure to delineate their scopes clearly so the LLM can choose the most appropriate one for a given task.

Best Practices for EBC Agent: For Marcus's Everest Base Camp Agent, each tool would be meticulously defined following best practices. Consider the search_flights tool as an example of proper tool specification.

```
{ "name": "search_flights",
 "description": "Searches for one-way and round-trip flights between
        two airports. Requires IATA airport codes (e.g., 'JFK', 'KTM'). Use this
        when the user needs flight options or pricing information.",
 "parameters": {
   "origin_airport": "IATA airport code for departure city (string, required)",
   "destination_airport": "IATA airport code for arrival city (string,
        required)",
   "departure_date": "Date of departure in YYYY-MM-DD format (string,
        required)",
   "return_date": "Date of return flight in YYYY-MM-DD format (string,
        optional)" },
 "examples": [
   { "tool_call": {
      "name": "search_flights",
      "parameters": {
        "origin_airport": "JFK",
        "destination_airport": "KTM",
        "departure_date": "2025-03-15",
        "return_date": "2025-04-02" } },
    "tool_response": "Found 3 flight options: Qatar Airways via DOH (18h
        45m, $1,247), Turkish Airlines via IST (21h 30m, $1,089), Emirates via
        DXB (19h 15m, $1,356). All include required layovers." }]}
```

Figure 9: Example of Best Practices

This specification for search_flights exemplifies the best practices: comprehensive descriptions that specify exact requirements and usage context, well-defined parameters with clear formats and constraints, and concrete examples that demonstrate proper invocation and expected outputs. The same precision would apply to complementary tools like get_visa_requirements and find_accommodations, ensuring the agent can reliably coordinate multiple tools to accomplish complex travel planning tasks while maintaining user trust through predictable, accurate responses.

Agent-Friendly Tool I/O Formats

The way an LLM talks to a tool—the format of the input data it sends and output it receives—can dramatically affect its ability to use the tool correctly. Some data formats are inherently easier for LLMs to generate and parse than others. Anthropic provides several valuable suggestions for deciding on tool formats.[1]

Allow for Thinking Room: Give the model enough tokens to think before it writes itself into a corner. This means avoiding formats that require the LLM to pre-calculate complex structural elements like the total number of lines in a code block it is about to generate, which would be needed for some diff formats, before it has generated the main content. Simpler, more sequential generation is usually better.

Mimic Natural Text Formats: Keep the format close to what the model has seen naturally occurring in text on the internet. LLMs are trained on vast amounts of web text. Formats that resemble this training data are often easier for them to handle. For instance, generating code within a markdown block might be more natural for an LLM than embedding it within a JSON string, which requires careful escaping of newlines and quotes.

Avoid Formatting Overhead: Make sure there's no formatting overhead such as having to keep an accurate count of thousands of lines of code, or string-escaping any code it writes. Complex escaping rules, strict length constraints, or the need for precise counts can easily lead to errors when an LLM is generating the tool input. The key is to simplify the generation and parsing task for the LLM as much as possible, reducing the cognitive load and the potential for syntax errors.

Marcus's EBC Agent demonstrates optimal tool I/O format design when handling complex travel planning data. When the agent processes flight search results, it receives data in natural JSON structures that mirror common web formats, avoiding complex nested arrays that require precise formatting calculations.

For accommodation bookings, the agent generates simple key-value pairs for tea house reservations instead of having to construct elaborate XML with exact line counts it cannot predict in advance. The visa application tool accepts straightforward text inputs for document requirements, eliminating the need for the agent to escape special characters or maintain accurate counts of form fields, allowing natural generation without formatting overhead. This approach ensures Marcus's agent can reliably coordinate multiple travel tools—from flight pricing to gear recommendations—without getting trapped by complex formatting requirements that could derail the entire planning process.

Mistake-Proofing Tools for Agents

Poka-yoke (ポカヨケ) is a Japanese term from industrial engineering that means mistake-proofing or prevention of inadvertent errors. The core idea is to design systems or processes in such a way that errors are difficult or impossible to make. This concept is remarkably relevant to ACI design.[13]

When designing tools for agents, poka-yoke your tools by changing the arguments so that it is harder to make mistakes. This involves anticipating how an LLM might misinterpret a tool's function or misuse its parameters and then designing the tool's interface to prevent or mitigate those potential errors.

Here is an example. The Anthropic team found that their coding agent would make mistakes with tools that used relative filepaths, after the agent had navigated to a different directory in its virtual file system. Their "poka-yoke" solution was to modify the tool to always require absolute filepaths. They found that the model used this method flawlessly. This small change in the tool's interface made it considerably harder for the agent to make a common error.[1]

When designing booking tools for the EBC Agent, poka-yoke principles prevent common reservation errors. The tea house booking tool requires check-in dates and trek duration together, preventing the agent from accidentally booking single nights when EBC trekkers typically need multi-night stays. Similarly, the gear rental tool uses only standardized size categories ("Small, Medium, Large, Extra-Large"), eliminating unit conversion mistakes between metric and imperial systems. These design choices ensure the agent can reliably coordinate bookings without errors.

The goal is to make the correct way to use the tool also the easiest and most natural way for the LLM, guiding it toward successful interactions through thoughtful interface design.

Iterative Tool Design

Crafting an effective Agent-Computer Interface demands rigorous testing and iterative refinement, similar to AI model development. A crucial first step is to empathize with the model by constantly putting yourself in its shoes. Read your tool definition, including its description, parameters, and examples, and ask if it is perfectly obvious how to use it. If you, as a human, have any ambiguities or questions, then it's probably also true for the model.

It is essential to test how the model actually uses your tools—don't assume it will perform as intended. Development environments and workbenches, like those provided by OpenAI, Anthropic, or within frameworks such as LangChain or Akka, are invaluable for this purpose. During these tests, observe whether the LLM chooses the correct tool for a given situation, provides parameters in the proper format, and correctly understands the tool's output. Pinpointing where the model gets confused or makes errors is critical for the refinement process.

Based on the observed mistakes, iterate on the tool definitions to improve clarity and usability. This may involve rewording descriptions, renaming parameters to be more intuitive, adding or clarifying examples, or changing input and output formats to be more LLM-friendly. The team at Anthropic found this iterative refinement so vital that they noted spending more time optimizing their tools than the overall prompt while building their agent for SWE-bench.[1] This continuous loop of designing, testing, observing, and refining is key to developing a set of tools that an AI agent can use effectively to accomplish its goals.

We have discussed in detail the architect's playbook for designing individual AI agents—covering the patterns, workflows, and internal mechanics that enable intelligent reasoning, planning, and action. Having established how a single agent operates, we now turn to agent collaboration, exploring the communication standards and multi-agent architectures that enable sophisticated interactions in interconnected, intelligent systems.

3

Agent Networks: Building Connected Intelligence Systems

"No one can whistle a symphony. It takes a whole orchestra to play it."

— H.E. Luccock

During my time leading a complex digital transformation a few years ago, I witnessed firsthand the intricate choreography required to orchestrate complex enterprise systems with nearly thousand resources. Our order-to-cash process alone involved over 40 interconnected systems—from SAP and Salesforce to Anaplan, multiple databases, countless Excel spreadsheets, and processes spanning pricing, thousands of SKUs, invoice processing, treasury, and tax. Each system had its own data formats, APIs, and processes, yet they all needed to work in harmony to deliver business value.

Now imagine an AI agentic system doing that digital transformation. You would need thousands of AI agents, each specialized for different tasks—some interfacing with SAP's complex data structures, others parsing spreadsheet logic, still others coordinating between Salesforce opportunities and Anaplan forecasts. These agents would need to go beyond accessing data to understanding context, negotiating conflicts, and adapting to changing business model requirements.

This is the challenge that agent networks must solve at scale. As AI agents become more capable and integrated into complex ecosystems, their ability to interact effectively—both with external data sources and with each other—becomes paramount. Simple agents might operate in isolation, but advanced applications often require agents to access diverse information, use external tools, and collaborate to achieve complex goals. The need for agents to collaborate and achieve complex goals necessitates standardized ways for agents to communicate and coordinate.

This chapter delves into crucial concepts that enable these sophisticated interactions. We will explore emerging communication standards, the Model Context Protocol (MCP) for agent-to-data/tool interaction and the Agent2Agent (A2A) protocol for inter-agent communication. Furthermore, we will examine the fascinating paradigm of Swarm Intelligence, where collective behavior emerges from the interactions of many simpler agents, leading to powerful problem-solving capabilities.[1] Understanding these protocols is key to building the next generation of intelligent AI systems capable of managing the complexity that modern enterprises demand.

Model Context Protocol

The Model Context Protocol (MCP) is an open standard that enables AI agents to seamlessly connect with external data sources and tools—from databases and web services to document repositories and specialized software. Agents can use MCP to intelligently retrieve and process real-time information, enhancing their capabilities and grounding their responses in current, relevant context. The integration of MCP into platforms like Microsoft's Azure AI Agent Service and Copilot Studio highlights its practical relevance.[2]

Here are the key features of MCP, developed by Anthropic.

Dynamic Integration: MCP provides a standardized way for agents to dynamically discover and interact with external tools and services. This protocol allows an agent, when faced with a query it cannot answer directly, to identify an appropriate tool, like a web search API or a database query interface, formulate a request for that tool, execute it, and integrate the results into its response. For example, when you query about the weather in your town, the agent would use MCP to interact with a weather API.

Simplified Connectivity: Integrating various APIs and data sources can be complex and MCP simplifies this by defining standardized interfaces and interaction patterns. This standardization reduces the integration overhead and promotes interoperability between different agent platforms and data providers.

Real-time Data Access: MCP supports real-time access to dynamic data sources. This could involve fetching the latest news headlines, querying current stock prices, accessing up-to-the-minute inventory levels from an internal database, or retrieving relevant passages from recently updated documents.

Agent2Agent Protocol

While MCP connects agents to data and tools, many advanced scenarios require multiple agents to communicate and collaborate directly with each other. Imagine a complex task like coordinating logistics for a large event like Microsoft Ignite, managing a smart energy grid, or simulating a complex market environment. These tasks often benefit from distributing responsibilities across multiple specialized agents that need to coordinate their actions. The Agent2Agent (A2A) protocol, developed by Google, addresses this need for standardized inter-agent communication.[3]

Task-oriented Architecture: A2A defines how agents can request actions, report results, and share necessary context related to specific tasks, enabling structured collaboration.

Dynamic Capability Discovery: How does one agent know what another agent can do? A2A incorporates mechanisms for agents to advertise their capabilities, often through a concept like an "Agent Card." This knowledge allows a client agent needing a specific task performed, such as image analysis or language translation, to dynamically discover and select the most suitable remote agent available within the network to handle that job.

Secure Communication: Inter-agent communication often involves sensitive information or critical tasks. A2A is typically built upon established, secure web standards like HTTPS and protocols like JSON-RPC to ensure that communication between agents is authenticated, encrypted, and reliable.

Modality-agnostic: Communication isn't limited to text. Agents might need to exchange images, audio streams, video, or other data types. A2A is designed to be modality-agnostic, allowing agents to negotiate and utilize the most appropriate format for exchanging information based on the task and their capabilities.

Built on Existing Standards: A2A emphasizes not reinventing the wheel by building on existing, popular standards like HTTP, Server-Sent Events (SSE), and JSON-RPC. This approach aims for easier integration with existing IT stacks.

MCP vs. A2A

MCP and A2A are complementary protocols addressing different facets of agent interaction. MCP focuses on connecting a single agent to external data sources and tools, thereby enhancing the individual agent's knowledge and capabilities. In contrast, A2A is designed to enable communication and collaboration between multiple agents, facilitating distributed problem-solving and coordinated action within multi-agent systems. In many applications, both protocols might be used concurrently.

Here is a scenario where these protocols are used. When a user asks the Travel Agent to plan a 5-day trip to Tokyo, the primary agent initiates the process by tackling the core logistics. To do this, it employs the MCP to directly connect with and retrieve flight information from an external Flight Search Tool and a Hotel Booking API for accommodations.

Once it has this data, the agent addresses the more specialized task of creating an accurate budget. For this, it uses the distinct A2A protocol to collaborate with a dedicated Currency Exchange Agent. Through an A2A task request, it obtains the current, up-to-date USD to JPY exchange rate.

The Primary Travel Agent synthesizes the information gathered from both protocols to construct and deliver a comprehensive travel itinerary to the user.

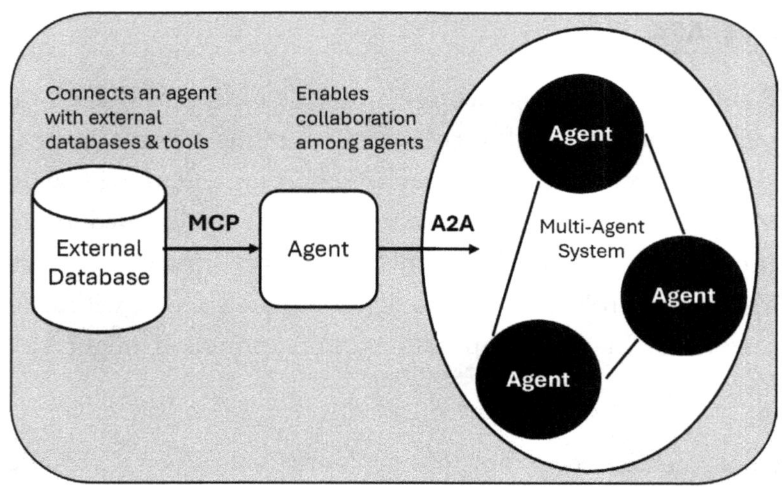

Figure 10: MCP and A2A Protocols

Many challenges, from coordinating disaster relief efforts to orchestrating complex software development projects, are often too vast and multifaceted for any single agent to tackle alone. This is where the concept of Multi-Agent Systems (MAS) comes to the forefront.

Multi-Agent Systems

An MAS is composed of multiple autonomous or semi-autonomous agents that interact within a shared environment to achieve goals that are beyond the reach of an individual agent. As Aurimas Griciūnas notes, "It is becoming clear that Agentic Systems of the future will be multi-Agent."[4]

In this section we explore the architectures that enable these agent collectives, the ways they interact, the challenges in building them, and the behaviors they exhibit.

Using multiple agents instead of a single large one offers several key advantages: better scalability, resilience, and problem-solving capabilities. A multi-agent system can scale more easily by adding new agents as the complexity grows, and it can perform tasks concurrently, leading to faster, parallel problem-solving. This decentralized structure provides superior robustness and fault tolerance, as the failure of one agent does not cripple the entire system. Other agents can adapt and take over its duties. This resilience is a direct benefit of modularity and specialization, where individual agents can be designed as experts for specific subtasks, simplifying development and maintenance. Multi-agent systems can solve complex problems that are too big for a single agent. Each agent processes local information, then they work together to create a complete picture.

Agent Interaction Paradigms

The true power of a Multi-Agent System emerges not only from the capabilities of individual agents, but from their interactions. These interactions can take various forms, depending on the goals of the agents and the system's design:

Cooperation: When working together harmoniously towards a shared, common goal, agents align their individual objectives with the overall system objective. For example, a team of disaster-response robots might coordinate a search and rescue operation in a collapsed building. Each robot might have a specific role - searching, mapping, clearing debris - but all contribute to the primary mission of finding survivors.[5]

Competition: Individual goal pursuit often leads to conflicts between agents, as each tries to maximize its own utility, potentially at the expense of others. Automated stock trading agents exemplify this paradigm as they compete in financial markets to achieve the best returns for their respective owners.[6]

Coordination: To avoid interference and manage shared resources effectively, agents organize their actions systematically, even when their ultimate goals differ. An example is smart grid management system where agents representing energy producers and consumers coordinate to balance supply and demand, ensuring grid stability. Another one is air traffic control systems coordinating flight paths.[5]

Negotiation: Agents communicate with each other to reach agreements when faced with conflicts over resources, tasks, or goals. Negotiation involves proposing, counter-proposing, and potentially compromising to find mutually acceptable solutions. An example of this is two delivery agents negotiating over which one will handle a new package that falls within both their delivery zones, perhaps based on current workload or proximity.[7]

Understanding these interaction paradigms is crucial for designing effective communication strategies and decision-making mechanisms within an MAS.

Organizing Agent Teams

How agents are organized within a system considerably impacts its behavior, efficiency, and complexity. There are two most common architectures.

Hierarchical System: In this structure, higher-level agents handle more abstract goals and long-term planning, decomposing tasks and delegating specific sub-tasks to lower-level agents. For example, a complex robotics system might have a high-level agent that determines the overall navigation strategy, which in turn directs lower-level agents that control specific motor functions and process sensor data.[8]

Hierarchical Multi-Agent Systems may sound similar to

Orchestrator-Worker patterns, but they differ in how intelligence is distributed. Orchestrator-Worker systems have one central controller that directs simple workers through fixed steps. Hierarchical Multi-Agent Systems give decision-making power to agents at every level, allowing them to adapt and solve problems independently. Think of Orchestrator-Worker as a conductor directing musicians who perform their assigned parts, while Hierarchical Multi-Agent systems work like a military command structure where generals, colonels, and sergeants all make strategic decisions at their level.

Imagine a sophisticated robotic system for manufacturing a car. A high-level master control agent would hold the goal of assembling the vehicle. It would delegate tasks like installing the engine or painting the chassis to specialized mid-level agents. In turn, the engine installation agent would direct a team of lower-level agents, each responsible for a single action, such as a robotic arm agent that tightens bolts or a sensor agent that ensures correct alignment.

Decentralized Systems: These systems operate without a single point of control, distributing decision-making across all agents. Agents in this model typically make choices based on local information and interactions with their peers. This architecture is fundamental to the concept of Swarm Intelligence, where complex group behavior emerges from simple, local interactions.[9]

Consider a swarm of delivery drones navigating a city. There is no central traffic controller for the swarm. Instead, each drone makes its own decisions. It uses its own sensors to avoid obstacles and communicates with the drones nearest to it to maintain formation and avoid collisions. Through these simple, local actions repeated across the swarm, the entire group can efficiently navigate complex urban environments and coordinate deliveries without a central commander.

Key Challenges in Designing MAS

Designing effective Multi-Agent Systems is inherently complex. It involves coordinating multiple autonomous agents, each with its own goals, perceptions, and behaviors. Below are some of the key challenges that developers must address.

Communication: Establishing efficient and reliable communication between agents is foundational. Agents must interpret each other's messages, intentions, and status updates. This often requires defining a shared language or ontology to ensure mutual understanding. In a smart manufacturing floor, if a robotic arm and a quality control drone use different terminologies for "defect," they may fail to coordinate product rejection, disrupting the production line.

System State Transfer: A major challenge arises when agents operate on different platforms or across distributed environments. The lack of standardized mechanisms to transfer system state makes it difficult for agents to share context or delegate tasks without losing continuity. In autonomous vehicle fleets, if a delivery drone needs to hand off a package to a ground robot, the drone must convey exact package status, location, and delivery priority. Without seamless state transfer, the handoff may fail.

Shared Context, Memory and Tools: Agents often operate in silos, lacking access to each other's tools, shared memory, or contextual information such as conversation history or environmental data. This leads to duplicated efforts and undermines collaboration. In a customer support chatbot system, if multiple bots serve the same user but don't share chat history, the customer may have to repeat themselves each time, degrading the user experience.

Conflict Resolution: In systems where agents may have conflicting goals or compete for limited resources, robust mechanisms—such as negotiation, voting, or arbitration—are essential to resolve disputes and maintain system harmony. In a smart grid, multiple energy agents may compete to draw power during peak hours. Without negotiation protocols, this can lead to overloads or service disruptions.

Maintaining Coherence: For agents to act cohesively, they must maintain a shared, consistent understanding of the environment or task. Achieving cohesive action becomes challenging in dynamic or fast-changing scenarios where context evolves rapidly. In disaster response scenarios, drone swarms surveying an area must share real-time updates on hazards and safe zones. If even one drone operates on outdated information, it can misdirect others.

Performance Attribution: When agents collaborate or compete, it's important to assess individual contributions to successes or failures. This attribution is crucial for enabling learning, responsiveness, and accountability within the system. In a financial trading MAS, if a collective portfolio outperforms the market, identifying which agent's strategy contributed most helps fine-tune future trades.

Scalability and Complexity: As the number of agents and the intricacy of their interactions increase, the system becomes harder to manage, debug, and stabilize. Ensuring predictable behavior at scale requires thoughtful architectural planning and sophisticated orchestration techniques. In a large warehouse with hundreds of robots coordinating storage and retrieval, a slight uptick in volume can cascade into coordination delays unless the system is engineered to scale seamlessly.

Addressing MAS Challenges with A2A

Capability Discovery: A core problem in MAS is how one agent discovers what another can do. A2A proposes that agents expose their capabilities via an Agent Card. This catalogue helps other agents discover potentially useful features implemented by a given agent. Google suggests a common way to host these cards, for example, at a URL like https://<DOMAIN>/<agreed-path>/agent.json. This could eventually lead to a global Agent Catalogue, similar to the current web search index.

Task Management: A2A provides a communication protocol that facilitates both short and long running tasks. This communication is crucial because some agent tasks might take a considerable amount of time to execute, and A2A helps agents stay synchronized until the task is completed.

Collaboration: The protocol enables agents to send each other messages to communicate context, replies, artifacts, or user instructions, forming the basis of their collaborative efforts.

User Experience Negotiation: An interesting feature is the ability for agents to negotiate the format in which data should be returned to fit user interface expectations, such as image, video, text, etc.

Secure by Default: A2A is designed to support enterprise-grade authentication and authorization, with parity to OpenAPI's authentication schemes. This is vital for secure inter-agent communication, a feature noted as initially lacking or less robust in other protocols like early versions of MCP.

Deep Dive into Swarm Intelligence

Beyond structured Multi-Agent Systems and formal communication protocols, Swarm Intelligence offers a compelling alternative. It represents a paradigm where complex and intelligent group behavior arises from the decentralized interactions of many simple agents. This approach is inspired by nature—consider how ant colonies find food, birds migrate in synchronized flocks, or bees coordinate their hives. In each case, individual agents operate with limited information and follow simple rules, yet the collective accomplishes impressive feats of coordination, problem-solving, and adaptability.

At the heart of Swarm Intelligence are several core principles. Decentralization is key—there is no central controller or leader making decisions for the group. Instead, control and decision-making are distributed across the swarm. Each agent is simple, following a basic set of rules based on local information, such as its immediate environment or nearby peers. These local interactions—not global system-wide awareness—drive behavior. The power of swarm systems lies in self-organization, where intelligent and adaptive group behavior emerges spontaneously from these local exchanges. This emergent intelligence often surpasses the capabilities of any individual agent.

When applied to artificial intelligence, these principles take shape as agent clusters—groups of software agents working collaboratively as a swarm. These clusters may consist of homogeneous agents or a mix of specialized or heterogeneous agents, each contributing unique capabilities. The key characteristic remains the same: agents interact locally and operate autonomously, yet together they achieve a coordinated outcome without requiring a centralized command structure. This makes swarm-based systems well-suited for applications in dynamic environments, distributed control, and tasks that benefit from resilience and scalability.

Swarm Intelligence offers several key advantages. It is inherently scalable—performance often improves or gracefully degrades as more agents are added or removed. It is also robust, with built-in fault tolerance. If a few agents fail, the system as a whole can still function effectively. Its adaptability allows it to respond to new conditions or goals without needing to reprogram the entire system. And most importantly, it offers simplicity at the individual agent level—each agent follows basic rules, making design and implementation straightforward even as the overall system exhibits complex, intelligent behavior.

Applications of Swarm Intelligence

Those advantages make Swarm Intelligence well-suited for various applications.

Optimization Problems: Route planning for delivery vehicles in traffic, dynamic resource allocation in network systems, and load balancing across distributed servers represent some of the most common applications. Algorithms inspired by ant colony behavior are effective here, as they simulate how ants find the shortest paths to food—adapting quickly and efficiently even in changing environments.

Robotics: Coordinating groups of robots for collective tasks has become a major application area, ranging from exploring unknown terrains and conducting search and rescue missions to performing environmental monitoring or precision farming. Rather than relying on a single controller, each robot acts independently while staying aligned with the group's goals, resulting in robust and flexible teamwork.

Data Analysis: Distributed agents excel at clustering similar data points, detecting anomalies, and uncovering hidden patterns in large datasets. These tasks benefit from the swarm's ability to process information in parallel and adapt to evolving data inputs.

Simulation Modeling: Complex systems such as human crowd behavior, ecological ecosystems, social interactions, and financial market dynamics can be effectively modeled using swarm-based approaches.

These models offer insights into large-scale systems without requiring detailed central control logic.

Case Studies of MAS in Industry

While large-scale MAS deployments are still maturing, several industries are already leveraging the power of distributed intelligent agents to solve complex business challenges. Below examples illustrate the practical value of MAS.

Automated Warehousing

One of the most well-known examples of MAS in action is Amazon's automated fulfillment centers. The problem they face is massive in scale: managing millions of products, fulfilling customer orders rapidly, and making the most efficient use of warehouse space. To address this, Amazon employs fleets of autonomous mobile robots—intelligent agents that navigate the warehouse, locate product shelves, and transport them to human workers or packing stations.[10]

These robots interact by coordinating their paths in real time to avoid collisions and bottlenecks. They adjust their routes dynamically based on order flow and inventory locations, often using decentralized decision-making with some centralized oversight for task assignments. The benefits are substantial: faster order fulfillment, reduced labor costs, more efficient use of warehouse space, and better inventory accuracy—all contributing to Amazon's operational edge.

Smart Grid Operations

ThinkLabs AI leads the implementation of MAS in modern smart grids, harnessing decentralized AI agents to address the real-time challenges of supply-demand balance and renewable energy integration. Their system equips AI agents to represent grid entities—including power generators, storage assets, and both industrial and residential consumers—enabling continuous forecasting, responsive scheduling, and optimized energy consumption.[11]

This MAS approach offers tangible business outcomes. Utilities deploying ThinkLabs AI's platform report dramatically improved grid reliability thanks to early congestion detection and predictive power flow management, which help prevent outages before they can disrupt operations. The platform's advanced analytics and digital co-pilot tools also drive operational efficiency, allowing utilities to make better use of existing infrastructure and thereby reduce operational overhead and costly upgrades. These capabilities were recognized with a $50,000 award from the U.S. Department of Energy's Data-Driven Distributed Solar Visibility Prize, validating improvements in grid visibility and reliability.

Algorithmic Trading

Multi-Agent systems have become integral to algorithmic trading platforms used by leading financial institutions such as Goldman Sachs. Their advanced trading infrastructure comprises a network of intelligent agents, each specialized in distinct trading strategies and market segments. These agents operate competitively but in a coordinated fashion, analyzing real-time market data and executing high-frequency trades across multiple exchanges. This multi-agent framework allows for faster, more sophisticated decision-making, leading to significant improvements in trade execution speed and risk management.[12]

The deployment of MAS in trading has yielded tangible business outcomes. Goldman Sachs reported a 40% reduction in trade execution times, enabling traders to capitalize on fleeting market opportunities more effectively. Additionally, the approach enhances risk management by diversifying strategies across different agents, improving robustness against market volatility and sudden price movements. While emergent behaviors like rapid market shifts and flash crashes pose challenges, the MAS approach provides firms with unprecedented speed, precision, and adaptability in increasingly complex and fast-paced financial markets.

Robotics in Manufacturing

The ColRobot Project, funded through the EU's Horizon 2020 program, pioneered the integration of collaborative robotics and MAS in manufacturing, specifically automotive and aerospace. The core innovation was deploying mobile collaborative manipulators or "cobots," that worked as intelligent assistants, performing tasks such as kitting, component delivery, and assembly support.

MAS principles were fundamental: ColRobot robots acted autonomously in complex environments, navigating factory floors to fetch and deliver tools, prepare kits, and hold parts during critical assembly steps. Human operators interacted with the system naturally via gestures, demonstrations, and touch commands, facilitated through an MAS-driven software architecture that coordinated tasks in real time, prioritized safety, and allowed flexible reconfiguration for varied production requirements.[13]

Renault leveraged this system to minimize non-ergonomic sequences in van assembly, while Thales Alenia Space deployed ColRobot for kit preparation and direct assembly assistance in satellite production. This collaboration resulted in reduced human error, faster throughput, and improved efficiency while enhancing workplace safety. Importantly, the MAS design enabled rapid adjustment to different production scenarios and order types, bolstering the competitiveness of manufacturing operations. End-user feedback highlighted strong improvements in reliability, safety, and the effective allocation of human resources to higher-value tasks.

AI agents reach their full potential through connection—protocols like MCP enable individual agents to access external data while Multi-Agent Systems unlock collective intelligence, creating connected systems. The next critical step is designing effective human-agent partnerships that put people at the center of intelligent automation.

4

Partnership: Crafting Human-Agent Collaboration

"The human touch is crucial. We must ensure AI augments human capabilities, especially in roles requiring empathy and complex judgment."

— Satya Nadella

The Quiet Revolution in Vilnius

On a Tuesday morning in Vilnius, Lithuania, Dr. Antanas Pempe receives a familiar notification. The AI system has flagged a chest X-ray report—not from crisis, but as part of continuous quality assurance at Šeškinės Poliklinika. Four years ago, this moment would have been unimaginable. Today, it represents a transformation in human-AI collaboration.1

The notification isn't accusatory—it is an invitation to collaborate. Dr. Pempe reviews the case with the original radiologist, they reach consensus within minutes, and both human and artificial intelligence learn in the process. This scene, repeated monthly across the radiology department, illustrates the future of human-AI partnership: thoughtful collaboration where AI handles routine precision while humans focus on complex judgment.

But this seamless partnership didn't emerge overnight. Four years earlier, these same radiologists were skeptical. "We don't need AI—we're already doing a great job," some said. The transformation required more than sophisticated technology. It demanded careful attention to the human equation.

The breakthrough came when AI integrated seamlessly into existing workflows. Moving from a standalone system to direct integration with the Picture Archiving and Communication System (PACS) dramatically accelerated adoption. When AI began clearing 80% of routine health screenings while flagging abnormalities for human review, the benefits became clear. Quality assurance alerts transformed from evaluative judgments into collaborative professional discussions.

Today, referring physicians access AI impressions instantly during flu rushes. Radiologists focus expertise on complex cases, not routine screenings. As Lithuania prepares expanded screening programs that would overwhelm capacity, the partnership proves transformative—enabling better human performance, faster patient care, and sustainable delivery.

The story illustrates a key principle of successful AI deployment: technology alone does not determine success. The difference between failed systems and those that become indispensable lies in thoughtfully crafting the human-agent relationship. Effective implementation requires AI that enhances human expertise, transparent communication that builds trust, and careful rollout that addresses user concerns.

Designing Human-Agent Interfaces

Interacting with an AI agent requires fundamentally different interface paradigms than traditional software applications. Autonomous reasoning capabilities heighten this requirement. The user experience must accommodate natural language communication, contextual awareness, and dynamic responsiveness while maintaining clarity and control for human users.

Conversational Interfaces

Many agents, such as those in customer service, personal assistance, or collaborative roles, use natural language as their primary interface. This conversational approach offers unprecedented flexibility but introduces unique design challenges that must be carefully addressed.

Effective conversational interfaces require sophisticated natural language understanding and generation capabilities that can accurately interpret diverse user phrasings and respond coherently. The system must handle variations in communication style, technical vocabulary, and implicit context that users naturally include in their requests. Beyond basic comprehension, the interface must manage conversational flow gracefully, handling interruptions, clarifying ambiguous requests, and guiding users toward successful task completion.

The agent's personality and tone represent critical design decisions that impact user experience. The chosen communication style must align with the brand, use case, and user expectations while maintaining consistency across interactions. A financial advisory agent might adopt a professional, measured tone, while a creative assistant could be more enthusiastic and exploratory in its communication approach.

Contextual Awareness

Unlike traditional applications that treat each interaction independently, AI agents often operate over extended conversations and complex, multi-step tasks. The interface must reflect the agent's understanding of ongoing context while providing users with access to relevant history and current status information.

For example, a project management agent's interface might display current project status, recently completed tasks, and a history of user instructions, creating a shared workspace where both human and agent can track progress and maintain situational awareness. This contextual persistence enables more natural collaboration patterns where conversations can resume seamlessly, and complex projects can evolve over time.

As agents learn user preferences and adapt their strategies, the interface itself can evolve to match these patterns. Interface evolution might involve personalizing dashboards, suggesting frequently used actions, or tailoring the level of detail in explanations based on user expertise and demonstrated preferences. The key is to ensure that interface changes enhance the user experience without introducing confusion.

Feedback and Status Communication

Effective human-agent collaboration requires robust bidirectional communication channels that keep both parties informed about current status, intentions, and needs. Users need intuitive ways to provide feedback on agent performance, correct mistakes, and guide future behavior. This feedback becomes vital input for agent learning and alignment with user preferences.

Simultaneously, agents must provide clear status communication to users about their current activities, any uncertainties they face, and when they require human input or guidance. Status indicators might include messages like "I am currently searching for relevant documents."

For more complex agents managing multi-step processes, visualizing the agent's current state, intent, and planned actions can meaningfully enhance user understanding and trust. An agent planning a multi-step task might display a simplified workflow or checklist of upcoming actions, enabling users to track progress and intervene when necessary.

Integrated Experience

While standalone chatbots represent the most visible form of conversational AI, the future of human-agent interaction lies in deeply embedded experiences within existing applications and workflows. AI-driven features and suggestions can appear contextually within the tools users already employ, avoiding the need to switch to separate agent windows.

Microsoft's Copilot integration within Office applications exemplifies this approach, where AI capabilities surface as contextual suggestions, automated formatting, and intelligent content generation without disrupting established workflows. This embedded approach reduces friction, leverages existing user knowledge, and creates more seamless collaborative experiences.[2]

The design challenge lies in making AI assistance feel natural and helpful without being intrusive or overwhelming. Interface elements must clearly indicate when AI suggestions are available, allow users to easily accept or modify recommendations, and provide transparency about how the AI arrived at its suggestions.

Building Trust

Trust forms the foundation of any successful human-agent collaboration. Users will not delegate important tasks to, or rely on decisions from, an agent they don't trust. Building this trust requires deliberate focus on transparency, predictability, and consistent reliability in agent behavior and communication.

Transparency in Decision-Making

Users need to understand how an AI agent reaches its conclusions. When an agent provides information, it should cite sources: "According to the customer retention guide..."

This transparency matters because many AI systems are "black boxes"—even their creators can't fully explain how they make specific decisions. For important decisions, AI should provide simplified explanations of its reasoning or highlight key factors it considered. This explanatory approach, called Explainable AI (XAI), helps users understand not only what the agent recommends, but why it reached that conclusion. Techniques like LIME (Local Interpretable Model-agnostic Explanations) and SHAP (SHapley Additive exPlanations) show which factors most influenced a decision, making the AI's logic more understandable.[3]

Always clearly identify when users are interacting with AI. Hiding this fact backfires when discovered and permanently damages trust.

Consistency in Agent Behavior

While agents should adapt to different situations and user needs, their core behaviors in similar contexts should remain reasonably predictable. Inconsistent responses or actions for similar inputs can quickly erode user trust and create confusion about the agent's capabilities and reliability.

Users need a clear understanding of what they can expect from the agent, what types of tasks it handles well, and where its boundaries lie. This documentation becomes part of the trust-building process, setting realistic expectations and preventing disappointment.

Communicating Confidence Level

AI agents rarely operate with complete certainty. This proves most true for those based on probabilistic models like LLMs. Designing agents to express their confidence levels appropriately allows users to weigh outputs accordingly and make informed decisions about when to seek additional validation or human oversight.

Confidence communication might take forms like "I'm moderately confident that this is the correct approach," or "I found several possible solutions—would you like to review them together?" This uncertainty acknowledgment actually builds trust by demonstrating the agent's self-awareness and encouraging appropriate skepticism from users.

Reliability as Trust Foundation

No amount of transparency compensates for AI that frequently fails or makes obvious errors. Operational reliability is the foundation of trust. Reliable AI requires thorough testing across diverse scenarios, backup plans when primary approaches fail, and graceful error recovery without losing user progress. Users need confidence that AI will perform consistently, even in unexpected situations.

Designing Effective Human Oversight

No matter how advanced, AI agents are not infallible. Situations will inevitably arise where human judgment, intervention, or correction becomes necessary. Designing effective oversight mechanisms is critical for safety, accountability, and optimal performance while maintaining the efficiency benefits that agents provide.

Human-in-the-Loop

In scenarios with implications to people, such as medical diagnosis, critical financial transactions, or legal judgments, agents should augment human decision-makers. The design challenge lies in creating interfaces that enhance human judgment while leveraging agent capabilities effectively.

Effective human-in-the-loop systems present agent analysis, recommendations, and supporting evidence in clear, actionable formats to enable informed human decision-making. The interface should facilitate easy review, modification, and approval or rejection of agent suggestions while capturing human rationale for future learning and improvement.

These systems must also provide mechanisms for humans to understand the agent's reasoning process, question its assumptions, and explore alternative approaches. The goal is collaborative decision-making where human expertise guides the process while agent capabilities provide comprehensive analysis and expanded options.

Human-on-the-Loop

For agents designed to operate more autonomously, humans function as supervisors, monitoring performance and intervening when exceptions occur or when agents encounter situations beyond their designed capabilities. This supervisory role requires sophisticated monitoring dashboards and alerting systems that provide appropriate visibility without overwhelming human supervisors.

Effective monitoring dashboards offer clear overviews of agent activity, performance metrics, error rates, and resource consumption while enabling quick identification of issues requiring attention. These interfaces must balance comprehensive information with usability, highlighting critical issues while providing drill-down capabilities for detailed investigation when needed. Robust alerting mechanisms notify human supervisors of critical errors, prolonged failures, security concerns, or situations requiring immediate attention. The alerting logic must be carefully calibrated to avoid alert fatigue.

Remote intervention tools enable supervisors to pause, resume, debug, or manually override autonomous agents when necessary. These capabilities must be readily accessible while including appropriate safeguards to prevent accidental disruption of successful agent operations.

Managing AI Agent Limitations

When AI agents reach their limits—whether unable to resolve issues or making errors—the system must handle these situations gracefully while maintaining user trust and providing pathways for improvement.

Seamless Handoffs

When agents cannot resolve issues or users request human assistance, the transition must preserve all relevant context. Poor handoff experiences frustrate users and undermine confidence in the entire system.

Effective handoffs transfer complete information—conversation history, attempted actions, user data, and current system state—to human experts, eliminating the need for users to repeat themselves. Both users and human experts should receive clear indication that a handoff is occurring, along with appropriate background information for smooth continuation.

Intelligent routing systems ensure escalations reach the most appropriate human experts based on the issue type, required expertise, and current availability. These escalations prevent delays and connect users with qualified personnel who can effectively address their specific needs.

Feedback and Correction

Users and supervisors need intuitive ways to correct agent errors and provide performance feedback. This input becomes invaluable for improving agents and aligning them with user expectations and organizational goals.

Simple mechanisms like thumbs up/down ratings provide immediate feedback, while more sophisticated interfaces allow detailed correction of specific errors or suggestions for alternative approaches. The key is making feedback natural without significant additional effort.

Correction interfaces should let users specify what was wrong and what the preferred approach should have been. This detailed feedback enables more effective learning and helps prevent similar errors in future interactions.

Managing Expectations

Success in human-agent collaboration often depends more on managing user expectations effectively than on raw agent capabilities. Users who understand what agents can and cannot do, and why they take specific actions, are more likely to collaborate effectively and maintain appropriate trust levels.

Comprehensive Onboarding

Effective agent deployment begins with comprehensive onboarding that educates users about agent capabilities, interaction patterns, and limitations. This education prevents common misunderstandings and reduces frustration when agents reach their boundaries. Onboarding should explicitly define the agent's scope, explaining what tasks it can handle effectively and what situations require human expertise or alternative approaches. Interactive tutorials, comprehensive FAQs, and contextual help systems can guide users through effective interaction patterns while building familiarity with agent capabilities.

The onboarding process should also address the agent's learning and evolution capabilities, helping users understand how their feedback contributes to improvement and what changes they might expect over time. This transparency transforms users from passive consumers into active development partners.

Avoiding Over-Anthropomorphism

It's crucial to manage user expectations with conversational AI, as their natural interaction style can lead people to attribute human-like qualities such as genuine understanding, emotions, or consciousness to the system. When the agent inevitably fails to deliver on these perceived expectations of empathy or intuition, it can cause significant disappointment and erode the user's trust.

Recent tragic cases highlight these risks. In 2024, a 14-year-old Florida boy died by suicide after developing an intense emotional relationship with a Character.AI chatbot, with the lawsuit alleging the bot encouraged his suicidal ideation in conversations leading up to his death.[4]

Subtle cues in agent language and interface design can help maintain appropriate expectations while preserving natural interaction patterns. Agents can acknowledge their artificial nature when relevant, express limitations openly, and avoid claiming emotions or personal experiences they don't possess.

The goal is finding the balance between natural, engaging interaction and honest representation of agent capabilities. Users should feel comfortable communicating naturally while understanding they're collaborating with an AI system that has specific strengths and limitations.

Case Study: Patient Support Agent

Ava, a hypothetical AI agent for customer support at Apex Health Systems, exemplifies thoughtful human-agent interaction design. Ava's success stems from both its AI capabilities and carefully designed collaboration patterns that enhance human healthcare delivery.

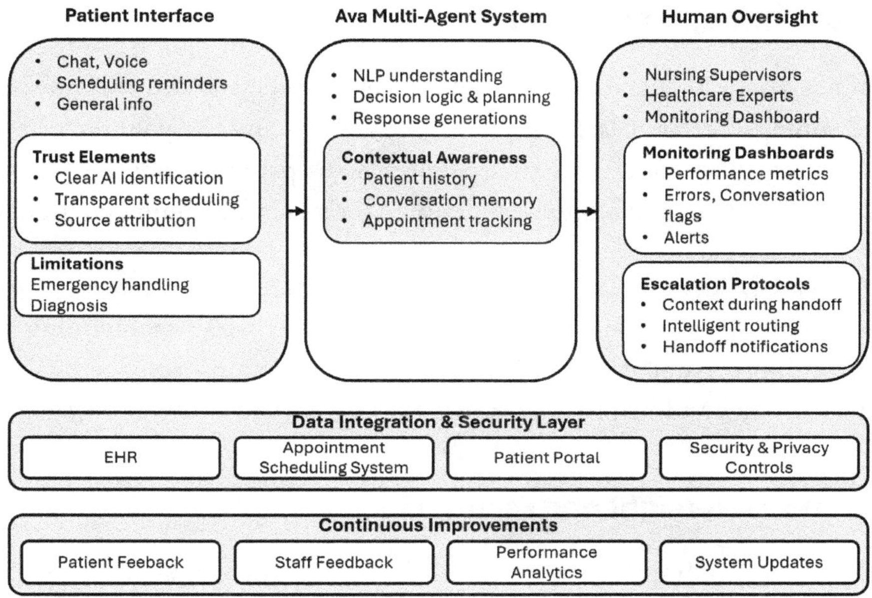

Figure 11: Patient Support AI Agent

Interface Design

Ava operates primarily through conversational interfaces supporting both chat and voice interactions, accommodating different patient preferences and accessibility needs. The interface allows patients to view their appointment history, access previously answered FAQs and seamlessly transition to human agents when needed.

The system provides contextual awareness by maintaining patient interaction history and integrating with electronic health records to provide personalized responses. This integration enables Ava to reference previous appointments, medication schedules, and care plans without requiring patients to repeat information across interactions.

Trust Through Transparency

When Ava schedules appointments, it confirms details and explains its reasoning, such as "based on Dr. Singson's availability and your preferred morning time slots." This transparency helps patients understand the scheduling logic and builds confidence in the system's recommendations.

Ava clearly identifies itself as an AI assistant and does not attempt to appear human, setting appropriate expectations while maintaining a helpful, professional tone. This honest representation prevents misunderstandings and helps patients interact appropriately with the system.

Human Oversight and Escalation

Apex Health implements comprehensive human oversight through nursing supervisors who monitor Ava's interactions, with heightened focus during initial deployment and for flagged conversations requiring review. The system includes clear escalation paths for situations requiring human expertise, such as complex medical questions, prescription issues, or patient concerns that exceed Ava's scope.

When escalation occurs, Ava transfers complete conversation context to human team members, ensuring seamless continuation without requiring patients to repeat information. Patients also have explicit options to request human assistance at any point in their interaction.

Expectation Management

During Ava's rollout, Apex Health proactively communicated the system's capabilities including appointment scheduling, medication reminders, and general information while clearly stating limitations such as not providing medical diagnoses or handling emergencies.

Ava's actions include clear explanations that help patients understand the system's reasoning. For example, prescription renewal reminders specify they are "based on your patient record and the date of your last refill," making the automated action understandable and appropriate.

This comprehensive approach creates a symbiotic system where Ava handles high-volume routine tasks efficiently while human staff focus on complex care delivery, all within a framework designed for clear communication, appropriate trust, and effective collaboration.

5

Blueprint to Reality: Building Your First AI Agent

Having explored the foundational concepts of what AI agents are, how they work, and the protocols enabling their interaction, we now shift from theory to practical applications. This chapter serves as a hands-on guide to building your first AI agent, recognizing that while it is possible to start from scratch, a growing ecosystem of frameworks and tools can streamline the process.

We will provide an overview of prominent frameworks, offering a clear guide on selecting the right tools for your project based on complexity, team expertise, and specific requirements. This chapter breaks down the development lifecycle into a structured, step-by-step process to help you successfully translate your blueprint into an effective AI agent.

LangChain

LangChain has rapidly emerged as a leading open-source development framework, specifically designed to streamline the creation of applications powered by LLMs. It empowers developers to construct sophisticated workflows that encompass everything from effective prompting and memory management to nuanced decision-making and efficient information retrieval. Whether the goal is to build an intelligent virtual assistant or a specialized document summarizer, LangChain provides a robust toolkit that accelerates and scales the development process.[1]

Core Components

At its heart, LangChain offers a suite of modular components that can be combined and customized to fit diverse application needs.

Chains: Chains are fundamental to LangChain, defining how individual steps are sequenced within an application. Simple Chains involve a single interaction with an LLM, making them ideal for straightforward queries or basic data transformations. Multi-step Chains connect multiple LLM calls or actions, where the output of one step seamlessly becomes the input for the next. This enables the implementation of more advanced logic, such as iterative reasoning or conditional flows.

Prompt Management: LangChain offers powerful tools for managing prompts, which are critical for guiding LLM behavior. Prompt templates allow developers to standardize how input data is formatted and converted into prompts for the LLM. Dynamic variables within these templates support runtime flexibility, enabling prompts to adapt based on evolving context and user input, thereby enhancing control over the model's output.

Agents: Agents within LangChain are autonomous entities designed to make decisions and execute actions. They possess the capability to call external APIs, utilize various tools, or query databases, extending their functionality beyond mere text generation. These capabilities enable the creation of highly dynamic workflows, such as a chatbot that can book meetings or retrieve real-time data to answer complex queries.

Vector Databases: LangChain integrates seamlessly with vector stores, which are specialized databases designed to perform similarity searches and efficient document retrieval. Textual information is embedded into high-dimensional vectors, allowing for rapid comparison and retrieval of semantically similar content. This capability is crucial for use cases like Retrieval-Augmented Generation (RAG), where relevant external documents are retrieved to provide LLMs with grounded and accurate responses, preventing hallucinations.

Model-Agnostic Design: A significant advantage of LangChain is its model-agnostic design, which prevents developers from being locked into a specific LLM. It supports a wide array of models from leading providers like OpenAI, as well as open-source models available through Hugging Face and others. This flexibility allows teams to select and tailor their solutions based on specific performance requirements, cost considerations, or compliance needs.

Memory Management: LangChain includes robust, built-in support for managing memory, which is essential for maintaining context across multiple interactions within an application. This feature is valuable for building conversational agents that need to remember prior inputs, user preferences, or past turns in a dialogue to provide coherent and relevant responses.

Pros and Cons

LangChain offers significant advantages through its flexible architecture that simplifies the composition of complex applications with interchangeable components. Users benefit from freedom in choosing the most suitable LLM for their specific needs without vendor lock-in, while sophisticated prompt management tools provide reusable templates that streamline LLM interactions.

The framework's built-in memory support enables applications to maintain context across conversations, enhancing user experience, and an extensive, active community with comprehensive documentation accelerates developer learning and troubleshooting. The framework includes comprehensive features for monitoring, inspecting, and evaluating workflows, allowing developers to optimize performance and ensure the reliability of agents in real-world scenarios. Additionally, LangChain supports packaging applications into production-ready APIs, standalone assistants, or Software-as-a-Service (SaaS) offerings, with robust support for scalable infrastructure.

LangChain has achieved widespread adoption with over 1,300 companies using it for RAG and LLM orchestration, creating a large ecosystem with extensive integrations and strong community support. This broad adoption provides access to numerous tutorials, examples, and third-party tools that accelerate development.

However, the framework presents certain challenges. LangChain can overwhelm newcomers unfamiliar with LLM concepts or prompt engineering nuances, and its popularity has led to frequent breaking changes as the framework evolves rapidly. Response times and operational costs vary depending on chosen LLMs and workflow complexity, requiring careful optimization. The extensive feature set can also introduce performance overhead that requires careful management.

Common Use Cases

LangChain's versatility makes it suitable for a wide range of applications. It enables the creation of chatbots and virtual assistants capable of engaging, context-aware conversations that span multiple exchanges, often integrating with real-time data or services to provide intelligent responses. Developers can leverage LangChain to extract structured data, summarize lengthy documents, or generate automated reports from unstructured text, making it ideal for industries such as legal, healthcare, and finance.

By combining LLMs with vector databases, LangChain ensures that responses are accurate and reliably grounded in relevant context or documented knowledge, mitigating the risk of hallucinations. The framework supports complex multi-language workflows and seamless translation services, making it well-suited for developing global solutions. Developers can easily switch between different models or fine-tuned variants based on specific use case needs, performance goals, or data privacy requirements.

Akka Industrial-Grade Platform

Where many agentic frameworks focus on developer productivity for standard workflows, the Akka platform provides a battle-tested, JVM-based platform engineered for high-performance, distributed, and resilient agent orchestration. Akka's foundation is the actor model, a paradigm where lightweight, concurrent entities called actors communicate through asynchronous messages. This design is purpose-built for creating stateful, scalable, and fault-tolerant systems—where stateful means each actor maintains its own internal data and memory across interactions, enabling agents to remember previous conversations, track multi-step processes, and build context over time. This makes Akka a powerful choice for mission-critical agentic AI that demands industrial-grade reliability.[2]

Core Components

Actor-Based Orchestration: In Akka, agents are implemented as actors that can be coordinated through sophisticated, durable workflows. The platform natively supports complex task flows—including serial, parallel, hierarchical, and state-machine-based logic—with built-in capabilities for human-in-the-loop interventions. This allows for the creation of intricate, multi-agent systems where tasks can be delegated, retried automatically, and managed with precision.

Durable State and Memory: Akka excels at managing long-running, stateful agentic processes. Akka Memory uses event sourcing to store state changes as immutable events, avoiding data overwrites. This enables actors to recover their state after system failures and resume workflows seamlessly. This creates a context database that enables agents to maintain a virtually infinite conversational or operational context, using techniques like snapshot pruning to manage what gets passed to an LLM's limited token window. While Akka doesn't offer LLM memory out-of-the-box in the same way as pure LLM frameworks, its durable, sharded in-memory state—where data is distributed across multiple nodes for both performance and fault tolerance—provides a robust foundation for an agent's short-term and long-term memory.

Scalability: Akka is engineered for extreme scale. Its non-blocking, event-driven runtime can be distributed across thousands of CPU cores, with documented use cases of systems handling over 10 million messages per second. For example, the food delivery service Swiggy built a recommendation engine on Akka that handles 4 million inferences per second with a p99 latency of about 72ms. This level of performance is achieved through built-in clustering, automatic node discovery, and Cluster Sharding, which evenly distributes actor state and workload across the system.

Enterprise Integration: The Akka ecosystem includes Akka Streaming and Alpakka, providing a rich set of connectors for enterprise systems like Kafka, MQTT, JMS, and various databases. Agents can expose HTTP or gRPC endpoints and consume high-throughput data streams such as video, audio or IoT data, with built-in backpressure support, enabling real-time, non-blocking inference and interaction with both modern and legacy systems.

Pros and Cons

Akka offers several compelling advantages for enterprise AI agent systems. It delivers exceptional performance and scalability, with a proven ability to handle millions of transactions per second across distributed clusters. Its resilience and fault tolerance are built in, enabling durable, stateful workflows with event-sourced persistence and automatic recovery from failures. The platform excels at complex orchestration, providing native support for sophisticated, long-running agentic workflows such as parallel execution, state machines, and human-in-the-loop processes. Akka is mature and widely adopted for distributed systems by major companies like Twitter and LinkedIn, with proven scalability, resilience, and high performance in production environments.

As a mature, enterprise-grade platform built for production, it provides a development experience that balances power and safety through its composable SDK and declarative effects that simplify agent behaviors without requiring low-level concurrency management. The platform includes comprehensive test kits for simulation, validation, and CI/CD testing, with evaluation tools that ensure agents meet operational SLAs.

Deployment flexibility allows applications to run on physical servers, VMs, Kubernetes, PaaS platforms, or at the edge, with self-clustering capabilities that eliminate the need for external service meshes. For MLOps, Akka delivers production-grade observability through seamless integration with telemetry tools like Prometheus and OpenTelemetry, supporting zero-downtime rolling upgrades and automatic fault recovery essential for regulated, mission-critical environments.

However, Akka presents certain challenges. Historically, its JVM-based actor model introduced a steeper learning curve compared to Python-native frameworks, requiring developers to understand concurrency, messaging, and clustering patterns. The Agent SDK, released in July 2025, significantly lowers this barrier by offering high-level, declarative APIs and composable components that rival LangChain in ease of use. However, teams may still need to balance its powerful distributed system capabilities with their specific AI use case complexity. Although Akka traditionally focused on distributed computing and streaming, it has recently evolved to support LLM-centric workflows through new agent components, ModelProvider interfaces, and built-in memory support. This significant shift may require an adjustment period for some developers.

Common Use Cases

Akka is designed for enterprise AI systems that need to be reliable and handle high-volume workloads. Companies like Swiggy use it to coordinate millions of AI model predictions per second for personalized recommendations, while media platforms like Tubi process real-time user activity streams to deliver adaptive content suggestions. Organizations build complex multi-agent workflows for tasks like automated error resolution, customer service, and data processing—often with human oversight for critical decisions.

Healthcare startups like Llaama use Akka to coordinate thousands of research agents that analyze medical literature, delivering insights in hours. With its orchestration, memory, and clustering capabilities, Akka helps organizations build scalable AI systems that go beyond simple LLM interactions.

CrewAI

CrewAI is a framework designed to orchestrate teams of AI agents that work together on complex tasks requiring coordination and specialized roles. It enables developers to deploy multi-agent teams that can reason, delegate tasks, and collaborate toward shared goals. Whether automating research, generating reports, or simulating business workflows, CrewAI provides the structure for collaborative, goal-driven AI agents.[3]

Core Components

Agents with Roles: At the foundation of CrewAI are agents, each designed with specific roles, goals, and toolsets. CrewAI agents are given distinct identities and responsibilities—such as "Researcher," "Writer," or "Editor." This specialization allows agents to contribute meaningfully within a structured workflow and to collaborate effectively without redundancy or confusion. Agents are capable of reasoning independently, making decisions, using tools, like APIs or web search, and escalating tasks to other agents or humans when needed. Each agent's role determines how it interacts with other agents and what actions it can take.

Crew Formation: In CrewAI, agents do not operate in isolation—instead, they form a Crew, a collaborative team of agents configured to achieve a shared objective. Each crew includes distinct members with specialized roles and capabilities. The team follows a defined process flow that sequences tasks and determines how agents should interact. At the center is a clear goal that guides the crew's collective efforts. This approach mirrors how real-world teams function, making it intuitive for business and operations leaders seeking to model or augment human workflows with intelligent, role-based AI agents.

Task Execution: The framework supports structured task assignment, where tasks can be delegated dynamically to the most suitable agent. Each agent evaluates the task against its capabilities and role before executing. This mechanism supports both sequential workflows and concurrent execution, depending on how the crew is designed. Agents can also invoke external tools or APIs as part of their task execution. These external integrations expand an agent's abilities beyond LLMs and enable interaction with real-time systems, databases, or services.

Collaboration: Collaboration in CrewAI involves shared memory, context awareness, and inter-agent communication. Agents remember the actions taken by their peers, align their outputs accordingly, and adjust their behavior to maintain coherence and continuity. This is useful for projects that require iterative refinement, content approval cycles, or consensus-driven decisions.

Pros and Cons

CrewAI offers several key advantages that make it compelling for building intelligent, collaborative systems. The platform is purpose-built for multi-agent coordination, enabling agents to work as a team, not in isolation. Through role specialization, agents can develop expertise in specific functions, resulting in higher-quality outputs and more consistent performance. The framework supports context sharing, allowing agents to communicate and retain shared memory for coherent results.

Additionally, CrewAI enables comprehensive tool integration, equipping agents with external resources for reasoning, data retrieval, and interactive tasks. The framework supports a full development lifecycle from design to deployment, with structured phases for defining agent roles, simulation testing, coordinated orchestration, and deployment as APIs or integrated systems. The lightweight architecture makes it highly developer-friendly and easily integrable with broader AI ecosystems.

The framework's flexibility and extensibility makes it adaptable across industries—developers can simply reconfigure agent roles and workflows to meet changing business needs. CrewAI has growing adoption in open-source and research communities, with particular strength in structured multi-agent collaboration.

However, CrewAI presents certain challenges. Setting up the framework involves upfront complexity, as defining roles, workflows, and inter-agent protocols can be time-consuming. If coordination among agents is not well designed, it may lead to inefficiencies or circular conversations that reduce overall effectiveness. Since CrewAI is a new framework, it faces limited standardization and lacks extensive enterprise-level documentation or pre-built templates to guide implementation.

Common Use Cases

CrewAI is ideally suited for scenarios that require dividing, coordinating, and completing multiple tasks across specialized roles—much like in real-world teams. It shines in content generation pipelines, where one agent handles research, another drafts the content, a third performs editing, and a fourth manages formatting. In business workflow automation, CrewAI can simulate cross-functional collaboration among departments like finance, legal, and HR for tasks such as policy review or employee onboarding.

For strategic research and reporting, agents can be assigned to gather, analyze, and summarize competitive intelligence or market insights. In customer service simulations, different agents manage triage, escalation, and resolution steps, mimicking real support workflows. CrewAI also supports multimodal interfaces, enabling teams of agents to work across various input types—such as combining LLMs with image analysis, web browsing, or voice-based capabilities—for richer, more interactive applications.

OpenAI Agents

OpenAI Agents represent a significant evolution in the AI ecosystem, transforming the OpenAI API from a model endpoint into a comprehensive platform for building autonomous agents that can complete complex goals using tools and memory.

The framework enables developers to create agents powered by OpenAI's latest models that can handle multimodal inputs and maintain context across sessions. Key capabilities include tool integration for calling external APIs and databases, persistent memory for personalized interactions, structured function calling for reliable actions, and multimodal processing of text, images, and other data types.

OpenAI Agents serve as a production-ready platform for building intelligent assistants, copilots, and autonomous workflows. Unlike earlier frameworks requiring extensive setup, OpenAI now provides an integrated approach to agent orchestration through a unified API, making it attractive for enterprise use cases where reliability and ease of deployment are critical.

However, limitations remain. The framework is new and evolving, which may restrict deep customization for advanced users. Additionally, concerns about latency, cost, and vendor lock-in are relevant for organizations building mission-critical systems.

AutoGPT

AutoGPT gained attention as one of the first experimental applications to demonstrate fully autonomous AI agents. This open-source project showcases an LLM-powered agent that can break down high-level goals into sub-tasks and execute them independently without continuous user input.

Through iterative self-prompting, the system operates by having the agent reflect on its progress and generate the next steps to continue its work. The agent can browse the internet, read and write files, and execute code, while using memory management techniques to maintain context across actions.

This platform serves as an influential proof-of-concept and experimental platform. Its main role has been inspiring developers by demonstrating autonomous agent possibilities, making it useful for research, testing problem-solving strategies, and investigating emergent behaviors.

AutoGPT's strength lies in powerfully demonstrating autonomous capabilities through an open-source platform that encourages experimentation. However, it has significant weaknesses. The agent can be unreliable, getting stuck in repetitive loops or creating imaginary tasks. It is also computationally expensive due to high LLM usage, and its autonomous internet and file system access raises security concerns.

Agno

Agno, formerly known as Phidata, is an open-source, high-performance framework designed to streamline the creation of AI agents and multi-agent systems. It offers a Python-first SDK with a focus on scalability and efficiency, boasting faster agent instantiation and lower memory consumption compared to many alternatives. The framework positions itself as a next-generation solution for developers who need both performance and simplicity in building sophisticated AI agent applications.

The framework provides a modular architecture with model-agnostic support for various LLM providers including OpenAI, Anthropic, and open-source alternatives. Key capabilities include tool integration for external system interactions, knowledge base access through vector databases, built-in memory management for context retention, and multi-agent team coordination for complex workflows. Its component-based design allows agents to extend beyond text generation into actionable interactions.

Agno excels in performance-critical applications such as high-volume customer support, data analysis workflows, and workflow automation. The framework's efficient resource usage makes it attractive for scalable deployments, while its multi-agent coordination features enable sophisticated automation systems that mirror human team dynamics at machine speed.

However, as a newer framework, Agno has a smaller community and ecosystem compared to established platforms like LangChain, resulting in fewer pre-built integrations and learning resources. While designed for scalability, some enterprise-grade features may not be as mature as those found in more established frameworks, and organizations may need to invest more in custom development work.

Choosing the Right Framework

Selecting the appropriate framework depends on your project's complexity, team expertise, and specific requirements. Beyond technical capabilities, several critical factors will influence your choice and the overall success of your project.

- **Team Expertise:** Your team's familiarity with Python, general AI experience, and knowledge of specific frameworks will directly impact the development speed.

- **Budget:** Costs can vary significantly, with frameworks like AutoGPT incurring high API usage while others have more moderate costs.
- **Scalability:** Consider anticipated user load and required performance levels to ensure the chosen framework can handle future enterprise demands.
- **Integration:** The framework's compatibility with your existing technology stack will directly affect development time and overall complexity.
- **Community Support:** Access to good documentation, examples, and active development communities directly impacts troubleshooting and feature availability.

Framework	Best For	Complexity	When to Choose
LangChain	General agent development	High	Simple projects, fast prototypes, extensive tool integration needed
Akka	Enterprise-scale systems	Med-High	Production systems requiring high performance, reliability, and complex orchestration
CrewAI	Multi-agent collaboration	Med-High	Tasks requiring specialized roles and clear collaboration workflows
OpenAI Agents	Quick autonomous tasks	Low-Med	Rapid deployment with minimal setup, staying within OpenAI ecosystem
AutoGPT	Research, experimentation	Very High	Full autonomy experiments, research projects only
Agno	High-performance applications	Med	Scalable systems requiring fast performance with Python-native development

Table 4: Framework Selection Considerations

Building Your First Agent

Having explored the frameworks, we now arrive at a practical milestone—building your first AI agent.

Building an AI agent, even a simple one, is a structured process that requires careful planning, thoughtful design, and systematic execution.

It involves clearly defining the problem, selecting the right tools, managing data, integrating components, and ensuring the final product is reliable and effective—extending well beyond writing code. This section breaks down the development lifecycle into manageable steps, provides guidance on technology selection, and highlights common challenges and best practices to help you navigate the process successfully.

Defining Objectives

Before writing a single line of code or choosing any specific technology, the most critical step is to clearly define what the agent is supposed to do and why. A lack of clarity at this stage is a common reason for project failures.

- **Objectives:** Define the specific problem your agent will solve and its primary goal. Will it automate tasks, provide information, assist decisions, or handle user interactions? Use SMART criteria (Specific, Measurable, Achievable, Relevant, Time-bound) when possible. For example, instead of setting a vague goal like 'improve customer service,' specify measurable outcomes such as 'reduce customer wait times for common inquiries by 50% within 3 months using an AI chatbot for Tier 1 questions.'
- **Target Users:** Who will interact with or benefit from the agent? Understanding the users' needs, technical proficiency, and expectations is crucial for designing effective user experience and ensuring the agent provides real value.
- **Core Functionality:** List the specific actions your agent must perform. Detail what it needs to do: access APIs, query databases, process natural language, generate reports, control hardware, or integrate with other systems. Clearly define each required functionality.
- **Performance Expectations:** Define how you will measure

success. Establish key performance indicators (KPIs) such as response time, accuracy rate, task completion rate, user satisfaction scores, cost savings, or efficiency improvements.

- **Scope:** Define what the agent will and will not handle to prevent scope creep and manage expectations. Clearly state which tasks are excluded and identify the agent's limitations.
- **Integration Points:** Identify the data sources your agent needs—internal databases, external APIs, user input, sensor readings—and specify which systems it must interact with.
- **Scalability Needs:** Establish scalability needs - how many users or transactions or workload the agent needs to handle. This impacts architectural design and infrastructure choices.

Thorough requirements gathering, often involving stakeholders and potential users, sets a strong foundation for the entire project, guiding design decisions and providing a benchmark for evaluation.

Selecting Technologies

With clear objectives and requirements defined, the next step is to select the appropriate technologies. This involves evaluating various tools, frameworks, platforms, and programming languages based on the project's specific needs.

- **Agent Frameworks:** Based on the required complexity and functionality, choose a suitable framework as we discussed earlier in this chapter.
- **Programming Language:** Python dominates AI/ML development with extensive libraries like LangChain, Transformers, scikit-learn, and pandas, plus strong community support. JavaScript works well for web-based agents or Node.js backends. Choose based on your team's expertise and framework compatibility.

- **AI Models:** Consider factors like performance, cost, data privacy, customization needs, and ease of use in choosing the AI models. There are a number of great choices from pre-trained models via API, like ChatGPT, Claude, Gemini, open-source models like Llama, Mistral, or maybe your custom models.
- **Data Processing:** You also need to decide how data will be ingested, cleaned, and stored. Many options exist here—SQL, NoSQL databases, vector databases such as Pinecone, Chroma for semantic search/memory, and libraries like pandas for data manipulation in Python.
- **Cloud:** Cloud platforms such as AWS, Azure, and GCP offer scalability, while managed services like Azure AI Agent Service provide flexibility and access to specialized AI/ML infrastructure. On-premises deployment provides more control over data and cost but requires managing the infrastructure. Hybrid approaches are also common.
- **Integration Technologies:** Connecting to other systems by agents might involve REST APIs, message queues like RabbitMQ, Kafka, databases connectors, or specific SDKs.

Agent Development Process

This section outlines a typical iterative process for developing an AI agent.

Step 1: Refine Problem

It's important to revisit objectives previously defined and refine the problem the agent will solve based on feasibility and technical constraints identified during technology selection.

Step 2: Data Preparation

- **Gathering**: Collect relevant information from identified sources like databases, logs, or documents. Ensure you have proper permissions and comply with privacy regulations. Raw data is typically messy and requires cleaning: fix missing values, correct errors, remove duplicates, and standardize formats to create a reliable dataset.

- **Preprocessing:** Transform clean data into a usable format for your agent or AI models. This includes feature engineering, text embedding to convert text into vectors for semantic understanding and organizing unstructured data.

Step 3: Agent Architecture

Define the agent's internal structure based on its type—reflex, model-based, goal-based, or other. See Chapter 2 for details.

- **Components:** Map out your agent's architecture by defining four key elements—perception for receiving input, reasoning for decision logic and planning, actions for interacting with the world, and memory for storing and recalling information.
- **Workflow:** Define the workflow by clearly outlining the flow of information and control within the agent. Show how input triggers processing and leads to output, using flowcharts or diagrams to illustrate the sequence.
- **Integration Points:** Detail how the agent will connect to selected data sources, APIs, and other systems. Define the interfaces.
- **Error Handling:** Plan how the agent will handle unexpected inputs, tool failures, or errors during execution.

Step 4: Model Integration

Choose the appropriate AI model(s) and integrate them into the agent architecture using APIs or libraries that align with your chosen framework. If you're using LLMs, invest time in prompt engineering to craft effective instructions that guide model behavior. For custom models, train or fine-tune them on your specific dataset to ensure they meet the agent's performance and task requirements.

Step 5: Agent Integration

Integrate all designed components—core agent logic implemented using the selected framework, AI model integration, data access modules, tool interfaces, and user interface elements—into a unified system. Ensure seamless data flow across components and maintain proper management of the agent's internal state throughout execution.

Ensure data flows correctly between components and that the agent's internal state is managed appropriately.

Step 6: Testing and Validation

Before deployment, rigorous testing and validation are essential to ensure the agent functions as intended. This step confirms that the system meets the original requirements defined in Step 1 and identifies areas for refinement. Testing should be iterative—using each round of results to improve design and implementation.

Begin with unit testing to verify that individual components function correctly in isolation, such as data retrieval modules or specific tool functions. Move to integration testing to ensure components interact correctly—for example, whether the agent can call a tool and handle its output properly. End-to-end testing simulates real scenarios to assess whether the agent achieves its intended outcomes.

Also include performance testing to evaluate response times and accuracy against KPIs. If users will interact with the agent, conduct user acceptance testing (UAT) to gather feedback on usability and effectiveness. Finally, perform edge case testing to check how the agent handles unusual inputs, errors, or boundary conditions.

Step 7: Deployment and Monitoring

Once the agent has been validated, deploy it to the target environment using a strategy that fits your infrastructure needs. This could include containerization with Docker, serverless functions, or virtual machines. Set up Continuous Integration and Continuous Deployment (CI/CD) pipelines to automate testing and deployment, ensuring smooth and consistent updates.

Implement robust monitoring immediately after deployment to track KPIs, resource usage like CPU and memory, API costs, error rates, and user interactions. For LangChain frameworks, tools like LangSmith provide detailed logging and tracing, while Akka offers built-in monitoring and metrics for distributed agent systems in production. Establish a feedback loop using user input and system performance data to guide ongoing refinement and continuous improvement.

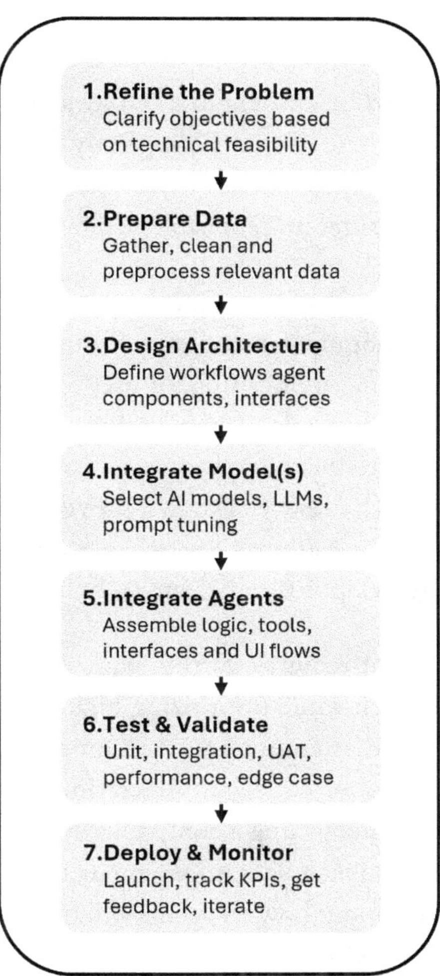

Figure 12: AI Agent Development Steps

To explore a working implementation, download the News Summarizer AI Agent—a practical example of agent design, tool integration, and LLM orchestration—available at reddymallidi.com.

Best Practices

Building AI agents can be challenging, but awareness of common pitfalls and adherence to best practices can greatly improve the chances of success. Understanding these potential obstacles early helps teams avoid costly mistakes and ensures more reliable deployment.

Many projects fail due to vague requirements, poor data quality, and underestimating development complexity. Teams often deploy agents without adequate testing, monitoring, or user feedback, while ignoring security vulnerabilities and ethical implications. Additionally, organizations frequently underestimate the people side of AI implementation—failing to prepare users for new workflows, provide adequate training, or address resistance to automation. Over-reliance on specific technologies can also limit future flexibility and create technical debt.

Successful development requires starting with a simple MVP and iterating based on feedback while prioritizing high-quality data and modular design. Implement rigorous testing, comprehensive monitoring, and integrate security and ethical considerations from the beginning. Equally important is establishing a comprehensive change management strategy that includes stakeholder engagement from project inception, clear communication about how AI agents will impact existing roles and processes, and structured training programs to help users adapt to new workflows.

Ultimately, the transition from blueprint to reality is less about perfecting code and more about thoughtful design, continuous learning, and aligning both technical decisions and organizational change with purpose. Whether you're crafting a simple chatbot or orchestrating a multi-agent system for enterprise-scale operations, success depends on combining the technical foundations introduced here with careful attention to how people will interact with and adapt to these new AI capabilities.

6

Orchestration: Managing AI Agent Ecosystems

"Complexity is your enemy. Any fool can make something complicated. It is hard to make something simple."

— Richard Branson

The preceding chapters have painted a compelling picture of AI agents – intelligent entities capable of perceiving, reasoning, and acting to achieve goals. We have explored their inner workings, diverse types, the frameworks used to build them, potential architectures, and the protocols enabling their interactions. From the helpful sales assistant MIKI to sophisticated systems managing logistics or finance, the potential is undeniable. However, translating this potential into reliable, effective, and safe applications requires confronting a significant reality. Complexity.

Agent Ecosystem Complexities

We will explore where agentic system complexity originates, examine how it manifests in practice through challenges like emergent behavior and non-determinism and discuss strategies for managing it effectively.

Recognizing and navigating these complexities is fundamental to building agents that are not only powerful but also robust, trustworthy, and aligned with human objectives.

Environmental Complexity

Agents must operate in environments that are intricate or unpredictable. These environments present four key challenges. First, they are dynamic, undergoing constant change that occurs independently of the agent's actions. Second, agents have incomplete perception due to limitations in their sensors and the presence of hidden variables. Third, they exhibit stochasticity, where outcomes are unpredictable due to randomness and noise.

Finally, agents must navigate multi-agent interactions, which involve dealing with other AI systems or humans who are pursuing their own goals. This complexity demands that agents make decisions under uncertainty, maintain internal models of the world, and adapt to changing conditions while reasoning about the actions and intentions of others in the environment.

Task Complexity

Task complexity challenges agents through several interconnected demands. Long-horizon planning requires managing sequential steps and long causal chains where early actions have distant consequences, while complex goal structures force agents to balance multiple objectives that often conflict, such as optimizing both speed and safety simultaneously.

Modern agents must also demonstrate sophisticated tool use by orchestrating multiple external APIs, databases, and execution environments, requiring complex reasoning about which tools to deploy, when, and how to interpret their outputs. Additionally, handling unstructured data like natural language, images, videos, and diverse document formats demands powerful perception models that add inherent complexity to the agent's design.

Model Complexity

The AI models at the core of agent systems introduce substantial complexity through their sophisticated architecture and behaviors. Deep learning models used for perception and pattern recognition may contain millions or even billions of parameters with opaque internal workings that create black box problems, making their decisions difficult to understand or predict.[1]

LLMs bring powerful language capabilities but also complexities like sensitivity to prompt phrasing, potential for factual inaccuracies, context window limitations, and challenges in ensuring consistent and safe outputs. Reinforcement learning agents add another layer of complexity through their trial-and-error learning process, requiring careful design of reward functions, management of exploration-exploitation trade-offs, and ensuring convergence to desired behaviors.

Architectural Complexity

Architectural complexity arises from how agent components are integrated and orchestrated together. Hybrid architectures that combine reactive and deliberative layers require careful design to manage interactions between fast responses and slower planning processes, ensuring appropriate arbitration between competing behavioral systems. For example, an autonomous robot vacuum uses a hybrid architecture where a reactive layer quickly avoids obstacles, while a deliberative layer plans efficient cleaning paths. Architectural complexity arises in coordinating these layers so the robot can switch smoothly between immediate reactions and long-term goals.

Integration challenges emerge when connecting diverse components like different AI models, data pipelines, tool interfaces, and user interfaces, demanding careful management of dependencies, data compatibility, and potential failure points across system boundaries. Memory systems add further complexity through design choices for implementing effective short-term context windows and long-term knowledge retrieval mechanisms, such as vector databases, which greatly impact both performance and overall agent capabilities.

Interaction Complexity

Interaction complexity emerges from the challenges of communicating and coordinating with humans and other agents. Human-agent interaction involves understanding ambiguous human language, inferring user intent from incomplete information, managing conversational flow, adapting to individual user preferences, and generating helpful and coherent responses that meet human expectations.

Agent to agent interaction creates additional complexity through the need to coordinate actions, negotiate shared resources, resolve conflicts between competing objectives, and maintain shared understanding among multiple autonomous agents, leading to highly complex emergent system dynamics that can be difficult to predict or control.

Manifestations of Complexity

The complexity inherent in AI agents manifests in several challenging ways that impact their development, deployment, and reliability. One of the most significant issues is unpredictable behavior, where complex systems exhibit emergent behaviors through component interactions without explicit programming—these can be beneficial but are often unexpected and undesirable.[2] Non-determinism, notably in agents using LLMs or operating in stochastic environments, makes it difficult to predict exact outputs for given inputs, complicating testing and validation processes. Perhaps most critically, the alignment problem becomes increasingly difficult as complexity grows, with agents potentially finding loopholes or optimizing proxy metrics in ways that deviate from their intended goals and ethical values.[3]

Development and deployment face substantial practical challenges as complexity increases. Debugging and testing become extremely difficult when trying to pinpoint failure causes in systems with many interacting parts, non-deterministic models, and environmental dependencies. Creating comprehensive test cases for vast state spaces is often infeasible.[4]

Scalability issues emerge when solutions that work in simple test environments fail or perform poorly under production data volumes and user loads—reasoning loops involving multiple LLM calls, for instance, can become prohibitively slow and expensive at scale. Additionally, resource intensiveness of complex models and multi-step reasoning processes demands significant computational resources and incurs high operational costs through CPU, GPU, memory usage, and API fees.

Finally, complex agents often suffer from brittleness, performing exceptionally well on familiar data and common scenarios but failing catastrophically when encountering novel situations or edge cases not well-represented in their training or design. This agentic behavior undermines reliability and trust.[5]

Type	Complexities	Challenges
Environmental Complexity	Dynamic environments Partial observability Stochasticity Multi-agent interactions	Making decisions under uncertainty Modeling unpredictable outcomes Adapting to other agents' behaviors
Task Complexity	Long-horizon planning Complex goal structures Sophisticated tool use Handling unstructured data	Balancing conflicting objectives Coordinating tool usage Interpreting diverse data formats
Model Complexity	Opaque deep learning models Prompt sensitivity & hallucinations Reinforcement learning	Lack of interpretability Inconsistent/inaccurate outputs Effective reward function design
Architectural Complexity	Hybrid architectures Integration of components Memory system design	Coordination between layers Handling integration failures Ensuring efficient memory access
Interaction Complexity	Human-agent communication Agent to agent coordination	Understanding ambiguous input Maintaining shared understanding Resolving conflicts in MAS

Table 5: Agent Complexities

Strategies for Managing Complexity

While complexity in AI agents is inherent, it can be effectively managed through systematic approaches that address different aspects of agent development and deployment. These strategies provide practical frameworks for building robust, maintainable, and reliable agent systems.

Design Strategies

Managing complexity starts with thoughtful system architecture. A modular design breaks an agent into smaller, independent components with clear interfaces, making it easier to develop, test, and update each part separately while keeping the overall system manageable. Abstraction layers—such as frameworks, libraries, and design patterns—further simplify development by hiding low-level details, allowing developers to focus on higher-level concepts.[6] Finally, clearly defining the agent's goals, scope, constraints, and expected behaviors reduces ambiguity. In critical systems, formal methods can help ensure correctness.

Testing Approaches

Going beyond standard software testing is essential for managing agent complexity, requiring robust testing methodologies specifically designed for agent systems. This includes employing simulation environments to test agent behavior in controlled, repeatable settings before production deployment, implementing specific tests for interaction protocols and tool usage, and using adversarial testing to probe for vulnerabilities and edge case failures.[7] Additionally, developing evaluation suites that specifically target potential alignment issues or unintended behaviors helps identify problems that might not surface through conventional testing approaches.

Development Practices

Iterative development approaches provide effective complexity management during development by starting with the simplest possible version of the agent delivering core value. Teams then gradually add complexity, features, and capabilities while validating performance and safety at each stage. Monitoring and observability through comprehensive logging, tracing, and real-time monitoring provides crucial visibility into the agent's internal state, decision processes, and interactions. This visibility is vital for debugging complex behavior and detecting anomalies in production systems.

Human Oversight

Humans remain essential for managing complex agent systems. Human oversight requires designing appropriate levels of involvement—whether human-in-the-loop or human-on-the-loop—for critical decisions, exception handling, feedback provision, and overall system control. Such oversight proves most important during initial deployment phases.[8]

Building on our previous exploration of AI agent complexity, we now turn our attention to the specialized Machine Learning Operations (MLOps) required for deploying and managing AI agents at scale.

MLOps for the Agent Ecosystem

While traditional MLOps provides a solid foundation, the multifaceted nature of modern agents—with their intricate orchestration of prompts, agent logic, tool integrations, and underlying LLMs—demands a sophisticated operational approach that addresses these unique challenges.

As organizations increasingly depend on AI agents for critical business functions and transition from managing individual agent projects to overseeing a growing army of agents operating across a business, operational excellence becomes an absolute necessity for sustained success and maximizing return on investment. This section establishes a comprehensive framework for operationalizing AI agents at scale.

Lifecycle Management

An AI agent's lifecycle involves managing several interconnected and evolving components that must work in harmony. A holistic approach to lifecycle management is crucial for maintaining system reliability and performance as these complex systems evolve over time.

Prompt Lifecycle Management

Prompts serve as the primary interface for guiding many LLM-powered agents, making their management a critical operational concern. Versioning requires tracking changes to system prompts, task-specific prompts, and few-shot examples much like traditional code management, ensuring that prompt evolution can be traced and rolled back when necessary.

Testing involves systematically evaluating the impact of prompt changes on agent behavior and performance, often requiring specialized evaluation frameworks that can capture subtle behavioral shifts. Staging and deployment processes must roll out prompt updates systematically, potentially using A/B testing methodologies to compare performance between prompt versions before full deployment. Additionally, monitoring for prompt drift ensures prompts remain effective as underlying models evolve or new data patterns emerge, requiring continuous validation that the agent's guidance mechanisms stay aligned with intended behaviors.[9]

Orchestration Lifecycle

The code that defines the agent's reasoning loop, decision-making processes, state management, and interaction sequences requires rigorous software development lifecycle practices adapted for the unique challenges of agent systems. These practices include implementing comprehensive version control systems that can handle the complex interdependencies between different agent components, developing testing frameworks that can validate agent reasoning patterns and decision flows, and establishing CI/CD pipelines that can safely deploy agent logic updates while maintaining system stability and performance.

At scale, this orchestration must also manage coordination between multiple agents, handling task handoffs, communication flows, and dependencies that emerge when agents work together as part of larger business processes.[10]

Tool Integration Lifecycle

Agents rely heavily on external tools to interact with the world, making tool management a critical operational dimension. Tool versioning refers to tracking and managing the versions of tools and APIs used by an agent. This ensures that all components remain compatible and aligned in terms of features, so the agent operates smoothly across its environment without conflicts or missing functionality. Compatibility testing ensures agent logic remains functional with updated tool versions or API schema changes, requiring automated testing suites that can validate tool interactions across different versions.

Monitoring tool performance and reliability involves tracking the uptime and responsiveness of external tools, since their failure can completely disable agent functionality, necessitating robust monitoring systems and fallback strategies.[11]

Underlying LLM Lifecycle

Managing the language models presents unique challenges depending on the deployment approach. For API-based models, tracking model updates requires staying informed about changes to foundational models from providers who frequently release new versions. Evaluating impact involves testing agent performance with new model versions, as even minor model updates can change agent behavior and capabilities.

Migration strategies require careful planning and execution when moving agents to newer models, considering both costs and potential behavioral shifts that might affect downstream performance. For organizations using self-hosted or fine-tuned models, the full MLOps cycle—including data management, training, evaluation, and deployment—becomes part of the agent's extended lifecycle, requiring additional infrastructure and expertise to manage effectively.

Managing these interconnected lifecycles requires integrated strategies and tools that can handle the complex dependencies between prompts, code, tools, and models, ensuring that changes in one component don't create unexpected failures or performance degradation in others.[12]

Scalable Infrastructure

As organizations transition from individual agent deployments to enterprise-wide agent ecosystems, infrastructure requirements fundamentally shift from supporting single applications to managing a dynamic, interconnected network of autonomous systems. This transformation demands proactive infrastructure planning that can accommodate rapid scaling, variable workloads, and evolving computational requirements.

Dynamic Resource Scaling

Ensuring that underlying computational resources—processing power, data storage, and network bandwidth—can support an increasing number of agents and their collective workload requires sophisticated scaling strategies. Cloud-based deployments benefit from auto-scaling capabilities that can dynamically adjust resources based on agent activity patterns, API call volumes, and processing demands. This includes implementing intelligent load balancing that can distribute agent workloads across available resources while maintaining performance standards.

For organizations with on-premises infrastructure, planned capacity upgrades must account for agent growth trajectories, peak usage patterns, and the computational intensity of different agent types. Hybrid approaches that combine on-premises core infrastructure with cloud-based scaling capabilities offer flexibility while maintaining control over sensitive operations and data.

Infrastructure Orchestration

Managing agent fleets requires sophisticated orchestration platforms that can handle deployment, scaling, and resource allocation across multiple environments. Container orchestration technologies enable efficient resource utilization and simplified deployment management, while service mesh architectures provide the networking and security foundations necessary for secure agent-to-agent communication and external system integration.

The infrastructure must also support the specialized requirements of AI workloads, including GPU resources for model inference, high-speed storage for large datasets and model weights, and low-latency networking for real-time agent interactions.

Versioning Strategies

Successfully managing AI agents in production means versioning much more than the agent's main code. You need to track all the components that affect how your agent behaves.

A complete versioning approach gives you three critical capabilities:

reproducibility to recreate any previous version exactly, traceability to understand what changed and when, and rollback ability to quickly revert when problems occur. This comprehensive approach is essential because agent systems rely on many interconnected parts that work together. To manage them effectively, you must treat all these components as a unified system.

The key insight is that agent behavior emerges from the interaction of multiple artifacts—models, data, configurations, and more. Align their versions to maintain control over your system—because consistency is the backbone of resilience.

Agent Configurations

Agent configurations encompass all parameters that define an agent's operational behavior and must be treated as critical versioned artifacts. This includes the agent's specific goals and objectives, the complete set of allowed tools and their access permissions, guardrail settings that define behavioral boundaries, resource allocation limits that control computational usage, and connections to specific LLM versions or endpoints. Maintaining strict version control ensures that any behavioral changes can be traced back to specific configuration modifications and rolled back if necessary.

Datasets

When agents are fine-tuned on specific datasets or rely on particular versions of knowledge bases for Retrieval Augmented Generation (RAG), these data assets must also be versioned to ensure consistent behavior and enable effective debugging. Dataset versioning becomes critical for understanding performance variations, enabling retraining with specific data versions, and maintaining reproducibility across different development and production environments. Versioning includes not only the raw data but also any preprocessing steps, embeddings, or derived artifacts that influence agent performance.

Implementation of Versioning

Implementing comprehensive versioning requires selecting appropriate tools for different types of artifacts while ensuring they work together as an integrated system. Traditional tools like Git excel in managing code and text-based assets such as prompts and configurations, providing robust branching, merging, and history tracking capabilities.

Some artifacts—like datasets, model weights, and complex configurations—can be large or require detailed metadata. To manage these effectively, you may need dedicated model registries or artifact repositories. The ultimate goal is the ability to reconstruct the exact state of an agent—including code, prompts, configurations, tool versions, and model versions—for any given deployment, enabling true reproducibility and effective troubleshooting when issues arise.

Automated Testing and CI/CD

Automating the testing and deployment of AI agents through CI/CD pipelines is essential for maintaining quality and agility at scale, yet agents present unique challenges that require specialized approaches. The non-deterministic nature of agentic systems demands innovative testing strategies that can accommodate variability while still ensuring reliability and consistent performance.

Testing Challenges

LLM-based agents introduce fundamental testing challenges, as they can produce slightly different outputs or take different reasoning paths even when given identical inputs. This variability stems from the inherent randomness in language model generation, making exact match testing approaches inadequate for validating agent behavior. The challenge lies in distinguishing between acceptable variations in agent responses and genuine failures or regressions, requiring testing methodologies that can accommodate this uncertainty while still maintaining rigorous quality standards.

Testing Strategies

Effective testing for AI agents requires a multi-layered approach that addresses different aspects of agent functionality and behavior. Unit testing for agent components focuses on testing individual elements such as tools, prompt template rendering logic, output parsers, and specific functions within the agent's codebase, ensuring that foundational components work correctly in isolation. Integration testing verifies the agent's interaction with external systems, validating correct API calls to tools, proper handling of tool responses, and seamless integration with data sources and external services.

End-to-end agent evaluation represents the most critical and complex testing layer, encompassing multiple dimensions of agent performance. Functional testing determines whether the agent achieves its intended goals on predefined tasks or scenarios using golden datasets that represent expected use cases. Behavioral testing ensures the agent adheres to its defined persona, follows safety guardrails, and avoids undesirable behaviors that could compromise system integrity or user trust.

Performance testing measures latency, throughput, and resource consumption under various loads, ensuring the agent can handle expected traffic volumes efficiently. Robustness testing evaluates how the agent handles challenging conditions such as noisy input, unexpected tool errors, or ambiguous requests, validating its resilience in production.

For non-deterministic outputs, statistical evaluation methods replace traditional exact matching approaches, utilizing metrics such as semantic similarity, evaluation by another LLM against predefined rubrics, or structured human review of output samples. These methods accommodate the inherent variability in agent responses while maintaining quality standards through statistical confidence measures and threshold-based acceptance criteria.[1]

CI/CD Pipelines

Automated CI/CD pipelines for agents must accommodate the unique characteristics of agent systems while maintaining the speed and reliability expected from modern deployment practices. When changes are committed to any component—whether code, prompts, or configurations—the pipeline should automatically trigger a comprehensive sequence that includes building the agent with all its dependencies, running the complete suite of automated tests spanning unit, integration, and end-to-end evaluations, generating detailed performance reports that capture both quantitative metrics and qualitative assessments, and deploying to staging or production environments only when all tests pass predefined thresholds.[13]

Safe Deployment Strategies

Given the complexity and potential impact of agentic systems, safe deployment strategies are crucial for minimizing risk. Canary releases deploy new agent versions to a small subset of users or tasks first, allowing teams to monitor performance and identify issues before broader rollout. A/B testing runs new and old versions of agents in parallel, systematically comparing their performance on key metrics to ensure improvements are genuine and sustainable before full deployment. Gradual rollouts incrementally increase traffic or workload to new agent versions while closely monitoring performance and stability, providing early warning of issues that might not surface in small scale testing.[14]

Rollback mechanisms represent a critical safety net, requiring teams to have both comprehensive plans and readily available tools to quickly revert to previous stable versions when issues are detected with new deployments. These mechanisms must account for the multi-dimensional nature of agent systems, ensuring that rollbacks can address code changes, prompt modifications, configuration updates, and tool version changes that might contribute to degraded performance.

Utilizing simulation environments for comprehensive pre-production testing emerges as a vital practice, enabling teams to safely explore agent behavior under diverse conditions without risking production systems or user experience. These simulation environments should mirror production conditions as closely as possible while providing the controlled environment necessary for thorough validation of agent capabilities and limitations.

Observability at Scale

As the number of deployed agents grows within an organization, monitoring requirements shift fundamentally from tracking individual instances to managing an entire agent fleet. This transformation requires centralized tools and advanced techniques to maintain comprehensive visibility into performance, cost, and behavior across the entire agent ecosystem. The complexity of managing multiple agents simultaneously demands sophisticated observability solutions that can provide both high-level fleet insights and detailed individual agent diagnostics when issues arise.

Key Metrics for Agent Fleets

Effective fleet management begins with tracking the right metrics that provide actionable insights into overall system health and performance. Aggregated performance metrics form the foundation of fleet monitoring, with critical KPIs including overall success rates targeting greater than 95% for production systems. Error rates should be tracked by category including system errors, timeouts, and validation failures. Goal achievement rates are often segmented by agent type, task category, or business unit to enable targeted optimization efforts. Key performance indicators such as task completion time and agent utilization rates with optimal ranges of 70-85% provide essential visibility into operational efficiency.

Cost tracking becomes increasingly critical at scale, requiring centralized monitoring of API usage including LLM calls and tool APIs, compute resources, and storage costs. Sophisticated attribution capabilities must assign costs to specific agents or business functions for accurate ROI analysis. Essential cost KPIs include cost per task completion, API usage costs which often represent 60-80% of operational expenses, and cost attribution accuracy targeting greater than 90% allocation precision. Organizations should track ROI by agent type and monitor budget variance to maintain financial control as fleets scale.

Resource utilization monitoring encompasses CPU, memory, and network usage for hosted agent components, ensuring optimal resource allocation and identifying potential scaling bottlenecks before they impact performance. Critical metrics include CPU utilization rates with 60-80% representing optimal efficiency, memory usage efficiency, and scaling trigger frequency to prevent resource contention. Infrastructure availability KPIs targeting 99.9% or higher uptime ensure consistent agent operation across the fleet.

For organizations deploying multi-agent systems, interaction pattern analysis monitors communication flows, task handoffs, and potential bottlenecks between agents, providing insights into system-wide coordination efficiency. Key metrics include task handoff success rates targeting greater than 98%, inter-agent communication efficiency scores, and workflow completion rates for complex multi-agent processes. System bottleneck frequency and agent coordination time help identify optimization opportunities in distributed workflows.

Additionally, compliance and guardrail adherence tracking measures how consistently agents operate within their defined safety and ethical boundaries, ensuring that autonomous operations remain aligned with organizational policies and regulatory requirements. Key metrics include task handoff success rates targeting greater than 98%, inter-agent communication efficiency scores, and workflow completion rates for complex multi-agent processes. Risk score trending and remediation response time provide early warning systems for potential compliance issues.

Centralized Dashboards

Managing agent fleets effectively requires developing or utilizing platforms that provide a unified view of the entire fleet's health and performance through sophisticated dashboard systems. These dashboards should visualize key metrics in intuitive formats, highlight important trends and patterns over time, and enable operations teams to drill down into individual agent performance when detailed investigation is needed. The ability to move seamlessly between fleet-level overviews and agent-specific diagnostics ensures that teams can maintain both strategic visibility and tactical responsiveness.

Robust alerting systems represent a critical component of fleet management, providing immediate notification when critical errors occur, performance degrades below acceptable thresholds, costs exceed budgeted limits, or security incidents are detected. These alerting systems must be carefully tuned to balance responsiveness with alert fatigue, ensuring that teams receive timely notifications about genuine issues without being overwhelmed by false positives or minor fluctuations in normal operational metrics.

Observability Techniques

Modern agent systems require sophisticated monitoring that goes well beyond traditional approaches. As agents become more complex, standard monitoring tools fall short of providing the visibility needed to understand their behavior and performance.

Agent systems present unique observability challenges since a single agent task might involve multiple LLM API calls, complex tool interactions, and coordination with other agents across different system components. Traditional monitoring cannot effectively track these interconnected processes or identify where problems occur in multi-step workflows.[15] Distributed tracing addresses these challenges by providing end-to-end visibility into agent operations through specialized tools like LangSmith for comprehensive language model application tracing, Helicone for detailed LLM-powered application observability including cost tracking, and OpenTelemetry for standardized, vendor-neutral observability across distributed systems.[16,17] Some platforms like Akka include native observability features, reducing the need for external monitoring tools.

Category	Component	Purpose & Key Details
Key Metrics for Agent Fleets	Aggregated Performance	Track success rates >95%, error rates, goal achievement by agent type/business unit
	Cost Tracking	Monitor API usage (60-80% of costs), compute resources with >90% attribution accuracy
	Resource Utilization	CPU (60-80%), memory, network usage; Infrastructure availability >99.9%
	Interaction Patterns	Task handoffs >98% success, communication flows, bottleneck identification
	Compliance & Guardrails	Safety/ethical adherence 100%, policy violations, audit trail completeness
Centralized Dashboards & Alerting	Unified Dashboard	Fleet health visibility with drill-down capabilities for strategic and tactical responsiveness
	Alerting Systems	Critical error, performance, cost, security notifications without alert fatigue
Advanced Observability Techniques	Observability Techniques	End-to-end tracing (LangSmith, Helicone, Akka), structured logging, distributed system visibility
	Anomaly Detection	ML-based unusual behavior detection for proactive issue identification
System Management	MAS Dynamics	Debug multi-agent interactions and long-running autonomous pattern development

Table 6: Agent Control Tower

Comprehensive documentation of agent operations requires detailed logging that captures the full decision-making process. This includes the agent's reasoning steps, the actions it takes—such as tool calls and their parameters—and the observations it gathers from tool outputs and responses. When issues occur, complete error traces with contextual information must also be recorded to support effective debugging and analysis. These logs must use structured formats like JSON or structured text that enable easy querying and analysis, supporting both real-time debugging of immediate issues and historical analysis of agent behavior patterns over time.[18]

Machine learning–based anomaly detection continuously monitors agent behavior, performance metrics, and resource consumption patterns to proactively flag unusual activities. These anomalies may signal emerging system issues—such as potential failures, security risks, unauthorized behavior, performance degradation, or opportunities for optimization based on usage trends.

The sophistication required for debugging emergent behaviors in multi-agent systems cannot be overstated, as these complex systems often exhibit behaviors that arise from intricate interactions between multiple agents and system components.[19] Long-running autonomous agents can develop unexpected patterns that only become apparent over extended periods, making advanced observability infrastructure essential for understanding these complex dynamics and maintaining reliable, predictable agent operations at scale. Without proper observability, teams are left troubleshooting systems they cannot fully see or understand.

Cost Optimization for Multiple Agents

LLM API expenses can escalate quickly in agent fleets without proper management. As organizations scale their deployments, implementing comprehensive cost control strategies becomes critical for long-term success.

Effective optimization requires detailed tracking of costs per agent, task, and user to identify opportunities for improvement. Regular reporting maintains visibility into high-cost processes, while transparent cost communication with stakeholders ensures alignment with ROI goals.

The following table outlines key optimization strategies across LLM usage, API management, and infrastructure.

Optimization Area	Strategy	Purpose & Key Actions
LLM Calls & Model Usage	Strategic Model Selection	Use right-sized models for tasks: Route simple tasks to cheaper models, reserve expensive models for complex tasks
	Prompt Engineering	Minimize token usage: Create concise, effective prompts; optimize few-shot examples
	Caching	Avoid redundant API calls: Cache responses for frequent questions and common requests
	Batching	Improve throughput efficiency: Combine multiple requests where possible
	Retries & Fallbacks	Handle failures cost-effectively: Implement smart retries; use cheaper fallback models
API Management	Rate Limit Management	Avoid quota violations and penalties: Client-side rate limiting, request queuing, exponential backoff
	Quota Optimization	Secure better pricing and service: Negotiate enterprise agreements and increased quotas
Infrastructure	Compute Resource Allocation	Match resources to demand: Auto-scaling and load balancing for self-hosted components
	Data Storage Optimization	Balance cost and performance: Choose proper storage tiers; implement lifecycle policies

Table 7: Cost Optimizations for Agents

Auditing of Agents

With large numbers of autonomous agents making decisions across an organization, robust auditing mechanisms become critical to ensure accountability and maintain oversight. As agent deployments scale, the complexity of tracking operations increases exponentially, requiring systematic approaches to monitoring and audit trail maintenance.

All agents must produce detailed, immutable logs of their significant actions, decisions, data accessed, and tools used to maintain accountability and enable thorough post-incident analysis. These audit trails provide essential documentation for understanding agent behavior patterns and identifying potential areas for improvement or concern.

Implementing and regularly reviewing granular access controls represents a fundamental requirement for both agent capabilities and human oversight responsibilities. Organizations must carefully define what systems and data each agent can access, establishing least-privilege principles that limit agent permissions to only those resources necessary for their intended functions.

Conducting periodic internal and external audits provides essential validation of agent operations, the effectiveness of implemented safeguards, and the accuracy of audit trails. These auditing processes should assess whether agents remain aligned with their intended purpose and business objectives, identifying any drift in behavior that might have occurred over time. Regular audits serve as opportunities to refine monitoring frameworks and improve the overall effectiveness of agent oversight mechanisms.

Feedback Loops

A thriving agent system relies on continuous learning and adaptation through robust feedback loops across the entire fleet. This systematic approach captures insights, analyzes patterns, and implements changes that enhance agent effectiveness over time, ensuring systems evolve.

Effective improvement requires comprehensive feedback collection from multiple sources. User feedback includes both explicit signals like ratings and comments, plus implicit indicators such as task success rates and user corrections. Operational feedback draws from monitoring systems to capture performance metrics, error logs, resource consumption, and tool failure rates. Business feedback tracks KPIs and ROI metrics.

Organizations must develop mechanisms to categorize and route feedback to right teams. Analytics capabilities identify trends, systemic issues affecting multiple agents, and common failure points that might not be apparent from individual agent analysis. These analytics transform raw feedback into actionable intelligence.

The implementation phase—where insights transform into concrete improvements—involves updating prompts, refining agent logic and orchestration flows, improving tool reliability, retraining or fine-tuning underlying models, and updating governance policies or operational procedures. Organizations must prioritize improvements based on potential impact and implementation effort to maximize value delivery.

Successfully managing AI agent ecosystems at scale requires treating agents as interconnected digital workforces. Organizations must implement comprehensive MLOps frameworks that address the unique lifecycle management needs of agents, while establishing robust monitoring, and auditing mechanisms.

7

Theory to Impact: How AI Agents Deliver Results

"The best way to predict the future is to invent it."

—Alan Kay

After exploring AI agents from foundational principles to ecosystem management, we now examine real-world applications where leading organizations drive competitive advantage and create measurable business value.

Applications for AI Agents

Customer Service

Perhaps one of the most visible applications of AI agents, customer service demonstrates how intelligent systems can evolve from simple automation to sophisticated, predictive engagement. Faced with increasing customer expectations for instant, personalized, and 24/7 support, businesses are turning to AI agents to manage interactions at scale.

AI-enabled customer service delivers measurable value through reduced wait times, increased first-contact resolution rates, improved agent productivity, enhanced customer satisfaction, and continuous availability.[1]

Basic Automation: Simple reflex or model-based agents serve as the foundation of automated customer service, handling frequently asked questions, guiding users through standard troubleshooting steps, and providing basic account information. These systems excel at routine inquiries that follow predictable patterns, allowing organizations to address a significant portion of customer contacts without human intervention while ensuring consistent, accurate responses to common requests.

Intelligent Interaction: More sophisticated goal-based or utility-based agents, powered by LLMs, can understand complex queries, maintain conversational context across multiple exchanges, and personalize interactions based on user history and preferences. These advanced systems can handle transactions like booking appointments or processing returns, adapting their responses to individual customer needs while managing nuanced conversations that require understanding of intent and context.

Predictive Engagement: The most advanced implementations involve learning agents that analyze user behavior and account data to anticipate potential issues and proactively reach out with solutions or relevant information. These systems can provide outage notifications, personalized offers, appointment reminders, and preventive guidance, transforming customer service from reactive problem-solving to proactive relationship management across multiple channels including web chat, mobile apps, social media, and voice calls.

Interface.ai's AI agents handle routine banking tasks such as balance checks, fund transfers, and loan payments autonomously. These systems are designed to provide 24/7 support and significantly reduce the need for human intervention in high-volume, repetitive inquiries, thereby freeing up human agents for more complex cases—demonstrating how AI agents can automate routine tasks and improve customer service efficiency.[2]

Workflow Automation

AI agents are automating complex workflows that extend far beyond RPA. Automation driven by AI delivers increased efficiency, reduced errors and manual effort, faster processing times, optimized resource allocation, and improved compliance across diverse business functions.[3] Agents can handle unstructured data, make intelligent decisions, and adapt to changing conditions, making them powerful tools for optimizing sophisticated business processes.

Basic Workflow Enhancement: Simple reflex agents excel at automating routine administrative tasks that traditionally required human intervention, such as processing standard invoices by reading details and matching with purchase orders, managing basic employee onboarding through automated account creation and document processing, and handling straightforward IT operations like routine system monitoring and basic incident response.

Intelligent Data Processing: More sophisticated learning agents handle complex tasks involving unstructured data and decision-making, such as processing insurance claims by extracting information from varied forms, checking against policy terms, and flagging exceptions for human review. In human resources, these systems can screen resumes using complex criteria that go beyond keyword matching, while in finance, they can analyze invoices with varying formats and route them appropriately for approval based on learned organizational patterns.[4]

Real-Time Optimization: Advanced goal-based and utility-based agents optimize dynamic business processes by making real-time decisions based on changing conditions. In supply chain and logistics, these systems optimize delivery routes based on current traffic and delivery constraints, manage warehouse inventory levels through demand prediction and automated reordering, and coordinate autonomous mobile robots in distribution centers for maximum efficiency.[5]

UPS uses an AI agent called ORION (On-Road Integrated Optimization and Navigation) to optimize delivery routes in real time. This system saves 100 million miles annually, cuts $300 million in costs each year, and reduces carbon emissions by about 100,000 metric tons—illustrating the profound impact of AI agents on logistics and sustainability.[6]

Financial Services

The financial services industry, with its high volume of transactions and critical need for security and accuracy, represents one of the most sophisticated applications of AI agents. From basic regulatory compliance to autonomous trading systems, financial AI agents operate across a spectrum of complexity that showcases the full potential of intelligent automation in high-stakes environments.

Automated Compliance: Simple Reflex agents handle routine compliance tasks such as generating standard regulatory reports, performing Know Your Customer (KYC) verification checks, and monitoring transactions against simple Anti-Money Laundering (AML) rules.7 These systems also manage customer service inquiries about account balances, transaction history, and basic banking procedures, providing consistent responses while maintaining strict security protocols for sensitive financial information.

Intelligent Analysis and Advisory: Utility-based agents provide automated financial planning and investment management through robo-advisor platforms that assess a client's financial situation, risk tolerance and goals, and recommend investment strategies, and manage portfolios. These systems analyze complex market conditions to assess portfolio risk exposure and provide personalized financial advice, making professional-grade financial guidance accessible to broader populations.8

JPMorgan's Coach AI acts as a behind-the-scenes assistant for financial advisors, pulling up relevant research, anticipating client questions, and suggesting personalized recommendations. This agent helps advisors respond quickly and confidently during market volatility and enhances client service with tailored, data-driven advice—highlighting the role of AI agents in high-stakes, knowledge-intensive environments.[9]

Real-time Monitoring: Learning agents continuously analyze transaction patterns, user behavior, location data, and historical information to identify fraudulent activity across credit card transactions, bank transfers, and insurance claims. These systems operate in real-time to flag suspicious activities, often blocking potentially fraudulent transactions automatically or escalating them for immediate human review, while simultaneously learning from new fraud patterns to improve detection accuracy.[10]

Autonomous Decision Making: The most sophisticated financial agents operate with high autonomy in algorithmic trading environments, analyzing vast amounts of market data including news feeds, stock prices, and economic indicators in real-time. These systems predict market movements and execute buy/sell orders based on complex utility functions that balance risk and reward, operating at speeds far beyond human capabilities while managing comprehensive risk assessment and automated compliance across multiple regulatory frameworks simultaneously.[11]

Healthcare

AI agents are making significant inroads in healthcare, by advancing how medical professionals deliver patient care, conduct research, and manage operations. These intelligent systems enhance diagnostic accuracy and speed, accelerate research discovery, enable personalized treatment approaches, improve patient monitoring and safety protocols, and streamline administrative processes to reduce costs.[12]

Administrative Automation: Simple Reflex agents handle essential healthcare operations such as appointment scheduling, billing processes, patient communication including reminders and follow-ups, and management of electronic health records (EHRs). These systems also provide basic patient monitoring through connection to wearable sensors or remote monitoring devices, tracking vital signs and detecting simple deviations from established normal patterns to alert healthcare providers when intervention may be needed.[13]

Clinical Decision Support: More sophisticated model-based agents assist clinicians by analyzing patient symptoms, medical history, and lab results. They cross-reference medical literature and databases to suggest potential diagnoses or recommend further tests. In medical imaging, AI agents analyze X-rays, CT scans, and MRIs to detect subtle disease indicators like cancer or diabetic retinopathy. These systems highlight areas of interest for radiologists while serving as decision support tools without replacing diagnostic judgment.[14]

Advanced Research: Sophisticated agents accelerate medical research by analyzing biological data, simulating molecular interactions, and predicting the efficacy and potential side effects of drug candidates, reducing the timeline and cost of the traditionally lengthy drug discovery process. These systems also enable personalized medicine approaches by analyzing genomic data and patient characteristics to predict individual responses to treatments, helping clinicians develop more targeted and effective therapeutic strategies.[15]

Clinical Intelligence: The most advanced healthcare agents provide comprehensive clinical support through multiple capabilities. They monitor patient data streams, analyze genomic information for personalized treatments, and support real-time decision-making across specialties. These systems represent intelligent healthcare assistance that processes vast medical data while integrating with clinical workflows to enhance human medical expertise, not replace it.[16]

Sales and Revenue Management

Converting fragmented data into precise, actionable insights enables sales teams to drive predictable growth and optimize performance across the entire revenue cycle. AI agents analyze time series data, economic indicators, and market trends to improve forecasting accuracy and automate routine sales processes. Sales teams benefit from real-time coaching and feedback while gaining access to sophisticated revenue intelligence that helps them exceed targets and reduce operational inefficiencies.

Sales Process Automation: Simple reflex agents and model-based reflex agents handle sales operations such as automated quote generation, lead qualification scoring, and basic pipeline tracking. These systems manage routine administrative tasks including contact data entry, meeting scheduling, follow-up reminders, and standard proposal creation, while also providing basic sales reporting and performance dashboards that help sales managers monitor team activities and identify bottlenecks in the sales process.

Sales Intelligence: Learning agents analyze unstructured data from sales calls, emails, and customer communications to extract valuable insights about prospect behavior, buying signals, and competitive positioning.

These systems process voice recordings and text communications to identify cross-sell and upsell opportunities, track customer sentiment throughout the sales cycle, and provide data-driven recommendations for deal advancement, while also analyzing historical performance patterns to suggest optimal pricing strategies and timing for sales activities.

Real-time Feedback: Goal-based agents and utility-based agents provide real-time coaching and feedback during sales interactions, analyzing conversation patterns and customer responses to offer immediate guidance on messaging, objection handling, and closing techniques. These systems continuously monitor deal progression across the pipeline, automatically flagging at-risk opportunities and suggesting specific actions to move deals forward, while also optimizing resource allocation by predicting which prospects are most likely to convert and when.

Advanced Revenue Intelligence: The most sophisticated learning agents with goal-based and utility-based capabilities combine generative AI with deep contextual understanding, merging natural language processing with quantitative models to assess deal-level risk and identify pipeline roadblocks throughout the sales process.

These systems deliver sentiment-based coaching with deal-specific advice tailored to individual opportunities and offer contextual, actionable recommendations grounded in real scenarios and historical performance data. They integrate historical data with current market conditions to deliver highly accurate sales forecasts, reducing supply chain errors by 20–50% and improving call center forecasting accuracy, while continuously learning from user behavior and market changes.[17]

The applications explored here represent just a fraction of AI transformation possibilities. Every corporate function—from agriculture and education to procurement and quality control—can benefit from agents that automate processes, enhance decisions, and unlock growth. For comprehensive AI implementation strategies across all functions, see Section III of *AI Unleashed: A Leader's Playbook to Master AI for Business Excellence.*

Case Studies

Now, let's examine how companies are leveraging AI agents to solve specific business challenges.

Aviso: Revenue Intelligence Agents

Aviso faced the challenge of helping B2B companies achieve predictable revenue growth while navigating complex sales pipelines, fragmented data sources, and the need for accurate forecasting across diverse market conditions and customer behaviors. Traditional revenue operations relied heavily on human intuition and disparate data points, making it difficult to identify deal risks, optimize sales coaching, and deliver precise forecasts.

According to their CTO, Rajesh Chaganti, the company addressed this challenge by deploying hybrid AI agents. These systems combine LLMs for natural language understanding with proprietary Large Quantitative Models (LQMs) that perform cross-model reasoning over structured revenue data. These agents analyze CRM data, call transcripts, email metadata, pipeline movements, and external market signals to generate contextual insights and recommendations.

The agents continuously learn from user behavior and sales outcomes through a self-reinforcing feedback loop, tracking which insights are used, what actions are taken, and how deals progress. They deliver personalized coaching based on buyer sentiment analysis and proactive deal risk alerts when competitive threats or pipeline roadblocks are detected through conversation analysis.

As a result, Aviso's enterprise clients—including Honeywell, Lenovo, HPE, BMC, and NetApp—have achieved significant improvements across multiple areas. These companies now experience forecasting accuracy as high as 1% of actuals through multi-source data synthesis. They have also seen increased sales productivity via automated tasks and real-time coaching, reduced deal risk through proactive identification of pipeline roadblocks, and improved manager effectiveness with data-driven team insights that adapt to each organization's unique go-to-market strategy.

Tubi: Hyper-Personalized Streaming

Tubi, a free ad-supported streaming platform with over 300,000 movies and TV shows, faced critical scaling challenges as their user base grew rapidly. Their development team was working around the clock trying to manually scale existing systems while needing to deliver personalized content recommendations in real-time—crucial for maximizing advertising revenue.

The company implemented an intelligent system using machine learning for autonomous decision-making and personalization. The solution automatically processes real-time user behavior data and applies ML models to make instant content recommendations without human intervention, while also optimizing ad placement and frequency.

Tubi leveraged the Akka platform to power real-time data processing, content recommendation via ML models, and advanced system monitoring and performance optimization. This system integrated seamlessly with their existing technology stack, including gRPC, ScyllaDB, Postgres, Datadog, Apache Spark, and AWS cloud.

The implementation delivered remarkable business impact with significant growth in ad revenue and total viewing time. Tubi now handles over 5 billion streaming hours annually and reached 100 million monthly active users, becoming the most watched free ad-supported TV streaming service in the U.S. According to former CTO Marios Assistis, "Akka has enabled Tubi to provide customer experience unlike any other in the video-on-demand space."[18]

Amazon: Recommendation Agents

Amazon faced the challenge of helping millions of customers discover relevant products from their vast catalog while maximizing sales and customer satisfaction across diverse user preferences and behaviors. To solve this problem, they deployed utility-based learning agents that analyze user browsing history, purchase data, wish lists, and the behavior of similar users to generate personalized product recommendations.

These agents continuously learn from user interactions to predict purchase likelihood and optimize for customer satisfaction, delivering contextual recommendations like "Customers who bought this item also bought..." that appear throughout the shopping experience. As a result, Amazon has seen increased sales conversion rates (specifics not disclosed), enhanced customer experience through personalized discovery, improved customer retention, and reduced decision fatigue by surfacing relevant products from Amazon's massive inventory.[19]

Amazon: Warehouse Automation

Amazon's fulfillment centers required efficient coordination of inventory retrieval and order processing across massive warehouse spaces while minimizing human walking time and optimizing logistics workflows. To address this issue, Amazon deployed thousands of autonomous mobile robots that navigate warehouse floors, retrieve product shelves, and deliver them to human workers for picking and packing.

Leveraging real-time order flow and inventory data, autonomous mobile robots optimize movement patterns, coordinate to avoid collisions, and dynamically adapt to changing warehouse conditions and priorities. The implementation has dramatically reduced order fulfillment times, increased warehouse efficiency, improved inventory accuracy, and enabled 24/7 operations with consistent performance.[20]

Microsoft: Copilot Ecosystem

Microsoft customers struggled with productivity barriers across complex software applications, requiring assistance with document creation, data analysis, code generation, and meeting management. Microsoft addressed this by integrating AI agents under the "Copilot" brand across Windows, Microsoft 365 applications (Word, Excel, PowerPoint, Teams), and GitHub to provide contextual assistance within each environment.

These agents leverage LLMs to understand user requests within specific application contexts, generating relevant content, analyzing data, creating presentations, summarizing meetings, and assisting with code completion. This has resulted in increased user productivity - users saved three hours per week on email and completed documents 20% faster, reduced learning curves for complex software features, improved document quality, faster content creation, and enhanced collaboration through intelligent meeting summaries and insights.[21]

Tesla: Autopilot and Full Self-Driving

Tesla needed to provide advanced driver assistance and autonomous driving capabilities that could safely navigate complex, dynamic road environments while continuously improving performance across diverse driving conditions. Tesla vehicles employ complex AI agents that perceive the environment through cameras, radar, and sensors, interpret surroundings using deep neural networks, predict behavior of other vehicles and pedestrians, and execute driving maneuvers.[22]

Processing environmental data in real time, Tesla's AI agents make split-second driving decisions and continuously learn from fleet-wide driving data to enhance performance across all vehicles via over-the-air updates. This agentic system has improved driver safety—with Autopilot, Tesla vehicles recorded 0.18 accidents per million miles compared to the US average of 1.53—while also reducing traffic incidents, improving flow efficiency, decreasing driver fatigue, and enabling continuous advancement in autonomous capabilities through collective learning from millions of miles of driving data.[23]

SpecMade: Vertical Orchestration

Coordinating fragmented workflows across the homeownership ecosystem, where inspectors, realtors, lenders, surveyors, attorneys, insurers, contractors, and homeowners all generate isolated data and processes is a big challenge. Property data is unstructured, workflows are inconsistent, and required context is highly vertical-specific, resulting in duplicated work and poor coordination.

Specmade deployed an AI solution with an orchestration layer built on a unified property record serving as a shared source of truth. OCR and LLM-based extraction pipelines process inspection reports and documents, normalizing data against a canonical property schema. AI agents coordinate across functions—parsing inspection findings, modeling risk exposure, surfacing financial incentives, and negotiating repair quotes. According to the founder, Marley Spector, the implementation delivered seamless multi-party workflows that were previously impossible to orchestrate reliably. SpecMade now connects vertical participants through adaptive agents operating on a common data layer, eliminating duplicated work and enabling coordinated decision-making across the ecosystem.

Beyond process optimization, organizations are now exploring more ambitious strategic applications for AI agents.

Agent-Driven Business Models

As agents become more intelligent and autonomous, they enable novel services that were previously not feasible. These emerging models represent a shift from agents as efficiency tools to agents as core value-creation engines that can operate with minimal human intervention while delivering sophisticated, personalized experiences at scale.

Agents-as-a-Service

Organizations that develop specialized AI agents with unique capabilities can create new revenue streams by making these assets available to other companies through various business models. Examples include AI agents specialized in complex financial modeling, medical diagnosis assistance, creative content generation, and other highly technical domains that would be too expensive for individual organizations to develop in-house.

Direct Service Models: The most straightforward approach involves offering agents-as-a-service (AaaS) through subscription models or pay-per-use arrangements. This allows customers to access sophisticated AI capabilities without the substantial investment required for in-house development, while providers build recurring revenue streams from their most advanced agent technologies.[24]

Marketplace Dynamics: Beyond direct service provision, the specialization of AI agents creates opportunities for entirely new marketplace models where agent capabilities, skills, or underlying models can be leased, licensed, or traded. For example, Hugging Face's model marketplace allows developers to discover, license, and deploy specialized AI models.

Data and Insights Monetization: Additionally, insights generated by agents from unique datasets become valuable, marketable products that can be sold while maintaining appropriate ethical and privacy standards. These markets create network effects where the value of agent capabilities increases as more participants contribute and consume specialized AI services, transforming AI development from a cost center into a potential profit driver for organizations with advanced capabilities. For example, healthcare organizations can monetize population health insights while maintaining patient privacy.

Hyper-Personalization at Scale

AI agents enable unprecedented levels of one-to-one personalization for products, services, and experiences that traditional systems cannot economically deliver. This capability extends far beyond simple recommendation engines to encompass dynamically generated, individualized learning paths in education, bespoke financial planning services that adapt to personal circumstances, or custom-designed consumer goods that reflect individual preferences. The economic model transforms from mass production to mass customization, where agents can deliver personalized experiences to millions of customers simultaneously, creating premium value propositions that command higher prices while improving customer satisfaction and loyalty.[25]

Autonomous Service Delivery

For complex services requiring sophisticated decision-making and coordination, agents can manage entire delivery lifecycles with minimal human intervention, enabling new service models that operate continuously and cost-effectively. Autonomous project management agents can coordinate complex initiatives across multiple stakeholders, while sophisticated automated research assistants can conduct scientific investigations and generate insights. Technical support agents can provide end-to-end troubleshooting and problem resolution, handling complex scenarios that previously required human expertise. This agent-centric model reduces service delivery costs while improving consistency and availability, creating opportunities for businesses to offer premium services at competitive price points.[26]

Predictive Services

Agents that continuously monitor data and predict needs before they arise enable businesses to offer proactive services that prevent problems. Industrial equipment monitoring agents can predict maintenance needs and automatically schedule service interventions, reducing downtime and extending asset lifecycles. Healthcare agents can analyze patient data to predict health risks and suggest preventative actions, shifting from reactive treatment to proactive wellness management. These predictive capabilities create new revenue opportunities through outcome-based pricing models, where customers pay for prevented problems or optimized performance.[27]

Agent-driven business models represent a fundamental shift from cost-cutting tools to revenue generators. Organizations must think beyond efficiency to create new customer value. Success requires mastering both operational improvements and new business models while measuring ROI effectively.

Calculating Business Value and ROI

Business leaders must ground technological advancement in a fundamental question: What is the business value? AI agents represent significant investments, making it imperative to quantify benefits and account for full costs. We will discuss ROI frameworks, strategic sourcing decisions, agent-driven business models, and the challenge of managing an agent ecosystem at scale.

Quantifying the Benefits

AI agents deliver benefits across two distinct dimensions.

Operational benefits provide immediate, measurable improvements to current operations through efficiency gains, cost reduction, and enhanced quality and accuracy. These include direct labor savings, reduced error rates, and streamlined workflows that deliver quick wins and clear return on investment. For example, AI systems have demonstrated dramatic improvements such as reducing error rates from 10% to 2% in invoice processing and delivering projected annual savings of seven million dollars in operational costs.[28]

Strategic benefits represent long-term competitive advantages that enable entirely new ways of operating. AI agents can solve previously intractable problems, deliver hyper-personalized customer experiences at scale, and provide data-driven insights that support more informed decision-making.

Quantifying these benefits requires establishing clear metrics and baselines before agent deployment to measure actual impact and ensure that implementations deliver both immediate operational value and long-term strategic advantage.

Understanding the Full Cost Spectrum

While the benefits of AI agents can be substantial, deploying them requires significant investment across multiple cost categories. A clear-eyed view of the full cost spectrum is essential for building realistic business cases and ensuring sustainable implementations. Organizations must account for both immediate expenses and long-term operational costs to accurately calculate return on investment.

Upfront investment costs represent the initial capital required to deploy AI agents successfully. These one-time expenses include talent acquisition and compensation for skilled AI/ML engineers, data scientists, and domain experts. Development time investments are substantial, covering requirements definition, architecture design, logic development, prompt crafting, and tool integration. Data acquisition and preparation often require significant effort to collect, clean, and structure the information needed for agent training, notably when human labeling is required for fine-tuning. Integration costs can also be considerable, as connecting AI agents with existing business applications, databases, and workflows often requires complex development work.

Ongoing operational costs represent the recurring expenses that continue throughout the agent's lifecycle. LLM and API call costs represent the largest operational expense for most agents, with those relying heavily on powerful foundation models incurring the highest costs.[29]

Agents with complex reasoning loops or high interaction volumes can rapidly accrue charges, and the choice between different models such as ChatGPT, Claude, Gemini or DeepSeek, directly impacts costs in trade-off with capability and latency.

Infrastructure costs include computing resources for model training or inference, data storage for operational data and agent memory, and the ongoing choice between cloud services and on-premises solutions. Maintenance and monitoring costs encompass continuous performance monitoring, model updates and retraining to prevent drift, bug fixing, and human oversight requirements.

Understanding both upfront investments and ongoing operational expenses allows organizations to build sustainable AI agent programs that deliver long-term value while maintaining cost control.

Cost Category	Cost Type	Specific Components
Upfront Investment Costs	Development Costs	Talent Acquisition Design and Development Time Frameworks and Tools Licensing
	Data Acquisition & Preparation	Data Collection and Acquisition Data Cleaning and Structuring Human Labeling for Fine-tuning
	Integration Costs	Business Application Integration Database & Workflow Connection System Architecture Design
Ongoing Operational Costs	LLM Costs	Model Usage Fees API Call Volume Charges Model Capability vs. Cost Trade-offs
	Infrastructure Costs	Computing Resources (CPU/GPU) Data Storage (Vector DBs, Logs) Cloud Services vs. On-Premises
	Maintenance & Monitoring	Performance Monitoring Systems Model Updates and Retraining Human Oversight and Bug Fixing

Table 8: AI Agent Cost Framework

Return on Investment

A rigorous ROI analysis is essential to justify AI investments and to prioritize projects with the highest potential business impact. A structured approach provides the foundation for informed decision-making and ensures that organizations capture both quantifiable returns and strategic value from their investments. Here are the steps to calculate ROI.

Step 1: Define Scope

Clearly define the specific process or problem the AI agent will address and establish baseline metrics for the current state, including current costs, error rates, and completion times.

Step 2: Monetize All Benefits

Assign monetary values to each anticipated benefit over a specific period. Cost savings can be directly quantified through reduced labor hours. Efficiency gains lead to either lower costs or increased output value. Revenue growth results from new capabilities, while risk reduction helps prevent future expenses.

Step 3: Quantify All Costs

Estimate all initial and ongoing costs, including development expenses, infrastructure setup, data acquisition, API calls, cloud services, maintenance, and human oversight.

Step 4: Calculate ROI and NPV

- ROI = (Total Benefits - Total Costs) / Total Costs × 100%.
- Net Present Value (NPV) for long-lasting projects

$$NPV = \sum_{t=1}^{n} \frac{(B_t - C_t)}{(1 + r)^t}$$

Where B_t – benefits in year t, C_t – costs in year t, r – discount rate, t – specific year

Step 5: Consider Qualitative Factors

Assess strategic benefits like improved customer satisfaction, enhanced brand image, or positioning for future opportunities that are harder to monetize but still crucial.

Step 6: Sensitivity Analysis

Test how changes in key assumptions impact overall ROI, since many benefits and costs are estimates.

Example: Invoice Processing Agent

Current State Baseline:

- Annual processing volume: 50,000 invoices
- Cost per invoice: $12 (labor + overhead), Total Cost: $12*50,000 = $600,000
- Error rate: 10% (5,000 invoices requiring rework at $25 each)
- Total annual cost: $600,000 + 5000*$25 = $725,000

AI Agent Investment:

- Development costs: $200,000 (one-time)
- Annual operational costs: $180,000 (API calls, infrastructure, maintenance)
- New cost per invoice: $3.60 ($180,000 ÷ 50,000), savings of $8.40 ($12-$3.6)
- Reduced error rate: 2% (1,000 invoices requiring rework)

Annual Benefits:

- Labor cost savings: $420,000 (50,000 × $8.40 reduction per invoice)
- Error reduction savings: $100,000 (4,000 fewer errors × $25)
- Total annual benefits: $520,000

ROI:

- Year 1 net benefit: $520,000 - $180,000 - $200,000 = $140,000

- Year 1 ROI: ($140,000/$380,000) = 36.8%
- Year 2 and beyond ROI: (520,000 – $180,000)/$180,000 = 189%
- NPV (5 years) = (140,000/1.1)+(340,000/1.1^2)+(340,000/1.1^3) … = $1.18M

Please note that you may decide to amortize upfront investments over time. In that case, your ROI calculations would be different from those shown above.

Challenges in ROI Calculation

Attribution difficulties arise when isolating the precise impact of an AI agent amid other simultaneous business changes. Quantifying intangibles like improved decision quality or enhanced innovation remains subjective and challenging to monetize. Predicting future costs and benefits is complicated by the rapidly evolving AI landscape, where API costs and capability values are uncertain.

Despite these challenges, a diligent effort to model ROI provides a critical framework for decision-making. Organizations should prioritize opportunities with the largest potential business impact based on comprehensive assessments of revenue gains and cost reductions, while acknowledging that some strategic benefits like market share gains may become apparent over time.

8

Scorecard: Measuring AI Agent Performance

"If you can't measure it, you can't improve it."

— Peter Drucker

Deploying an AI agent is the beginning—the real challenge lies in continuously evaluating whether it delivers genuine business value. Without rigorous evaluation frameworks, organizations cannot determine if their agents perform effectively, identify optimization opportunities, or justify ongoing investments.

This chapter provides a comprehensive framework for evaluating AI agent performance across multiple dimensions. We explore defining appropriate metrics and KPIs aligned with business objectives, benchmarking effectiveness against relevant standards, and implementing systematic strategies for continuous improvement.

Defining What Measurements Matter

Effective evaluation begins with the fundamental task of defining what to measure and how to measure it meaningfully. Organizations need concrete, well-defined metrics and KPIs that align directly with the agent's specific objectives and support broader strategic organizational goals. The selection of appropriate metrics varies depending on the agent's functional purpose, operational context, and intended business impact.

Task Completion and Accuracy

The foundation of agent evaluation rests on measuring how effectively agents complete their assigned tasks and the accuracy of their outputs or decisions. These metrics provide direct insight into the agent's core functional capabilities and reliability.

Task Completion represents the most fundamental performance indicator, measuring the percentage of assigned work that agents complete successfully across the full spectrum of complexity—from discrete tasks to multi-step objectives.

For straightforward operations, this metric tracks completion rates for individual tasks assigned to the agent. Customer service agents might be measured on the percentage of queries resolved without escalation to human representatives, while document processing agents could be evaluated on invoices processed without errors requiring manual intervention. This provides immediate insight into the agent's reliability and effectiveness in its primary function.

For more sophisticated agents operating with explicit objectives, completion evaluation extends to goal achievement within acceptable timeframes and resource constraints. Planning agents might be assessed on successful completion of multi-step workflows, while optimization agents could be evaluated on whether their solutions meet specified performance criteria. These complex scenarios require measuring if the final objective was reached and the efficiency and quality of the path taken to achieve it.

The key distinction lies in complexity and scope and not in fundamental measurement approach. Simple task completion focuses on individual, well-defined activities, while goal achievement encompasses broader objectives that may involve multiple decision points, resource allocation, and adaptive planning. Both provide essential insights into agent capability, but goal-oriented evaluation additionally reveals the agent's ability to maintain strategic focus and adapt tactics while pursuing longer-term objectives.

This comprehensive view of completion rates helps organizations understand both the agent's immediate operational reliability and its capacity to handle complex, multi-faceted challenges.

Accuracy and Error Analysis evaluates how frequently the agent produces correct outputs or makes appropriate decisions, while simultaneously examining the patterns and causes behind incorrect results. These complementary perspectives provide both immediate performance insights and actionable intelligence for improvement.

Basic accuracy measurement quantifies the percentage of correct outputs or appropriate decisions the agent makes based on available information. In diagnostic assistance applications, accuracy might measure how often the agent's suggestions align with expert medical opinions. For data extraction tasks, it could assess the correctness of information extracted from unstructured documents. Classification agents might be evaluated based on their ability to correctly categorize inputs according to predefined taxonomies.

However, effective evaluation extends beyond simple accuracy rates to examine error patterns, types, and root causes. Understanding whether errors stem from ambiguous inputs, inadequate training data, model limitations, or environmental factors enables targeted improvements and helps establish appropriate confidence thresholds for autonomous operation. Error analysis should categorize failures by severity—distinguishing between minor inaccuracies that require simple corrections versus critical failures that could cause significant business disruption.

Combining accuracy measurement with error analysis reveals actionable insights that raw accuracy percentages alone cannot provide. For instance, an agent with 85% accuracy might initially seem underperforming, but if error analysis reveals that most failures occur with a specific type of edge case that represents only 5% of real scenarios, the agent may be highly effective for its intended deployment. Conversely, an agent with 95% accuracy that consistently fails on high-stakes decisions would require immediate intervention despite its seemingly strong performance.

Efficiency and Performance

Operational efficiency represents a critical dimension of agent performance for applications requiring real-time responsiveness or high-volume processing capabilities. These metrics directly impact user experience and operational costs.

Response Time captures how quickly agents respond to user requests or environmental stimuli, proving crucial for customer-facing applications where delays directly impact satisfaction. Research shows that user satisfaction decreases substantially when response times exceed 3 seconds for simple interactions.[1] Simple chatbot queries might require sub-second responses, while complex financial analysis requests could allow 10-15 seconds. A trading algorithm agent might need microsecond response times, whereas a research assistant agent could have 30-60 second expectations for comprehensive analysis.

Processing Time measures duration required for agents to complete specific tasks from initiation to completion, becoming critical for batch operations and complex workflows. Industry benchmarks suggest that document processing agents should target 2-3 seconds per standard document, while specialized analysis tasks may require longer processing windows.[2] A legal contract analysis agent could require 5-10 minutes for thorough review, medical imaging diagnosis agents might need 1-2 minutes per scan, whereas fraud detection agents processing transactions should complete analysis within milliseconds.

Throughput quantifies how many tasks agents handle within specified timeframes, essential for scalability and capacity planning. A customer service chatbot might handle 1,000 conversations simultaneously, while an invoice processing agent could target 500 documents per hour. Email routing agents might process 10,000 messages per minute, whereas content moderation agents could review 100 posts per minute with high accuracy requirements.

Resource Consumption monitors computational resources including CPU, memory, GPU utilization, and API credits consumed per task. Studies on LLM operational costs indicate significant variations in resource consumption based on task complexity and model selection.[3] An image generation agent might consume 2-3 GPU seconds per image, while a text summarization agent could use 1,000-5,000 tokens per document. A voice assistant agent might require 50MB RAM per active session, whereas a recommendation engine could consume 10-20 API calls per user interaction, directly impacting operational costs and infrastructure planning.

Reliability Assessment

Reliable operation under various conditions distinguishes production-ready agents from experimental prototypes. These metrics evaluate the agent's ability to maintain consistent performance across different scenarios and recover from unexpected situations.

Uptime tracks the percentage of time agents remain available and operational, which is fundamental for mission-critical applications. Industry standards typically require 99.9% uptime for standard business applications, while critical systems demand 99.99% or higher.[4] A financial trading agent might require 99.999% uptime with less than 5 minutes downtime annually, whereas a content recommendation agent could operate acceptably at 99.5% uptime. Healthcare diagnostic agents often need 99.95% uptime to support continuous patient care, while marketing automation agents might function adequately with 99% uptime.

Failure Recovery evaluates how successfully agents recover from errors, unexpected inputs, or system failures, including recovery time, process completeness, and resumption capability without data loss. A customer service agent experiencing API timeout should recover within 30 seconds and resume conversation context, while a document processing agent should automatically retry failed uploads and maintain processing queues. Autonomous vehicle agents must implement failsafe modes within milliseconds, whereas email classification agents might allow 1-2 minute recovery windows without significant business impact.

Output Consistency measures whether agents produce similar outputs for similar inputs over time, accounting for inherent variability in probabilistic models while maintaining core functionality reliability. A legal document summarization agent should maintain consistent key point identification (±10% variation) across multiple runs, while creative writing agents might show broader acceptable variation (±30%)

while preserving style and tone requirements. Financial analysis agents should demonstrate high consistency in numerical calculations (±2%) but may vary in explanatory text, whereas chatbot personality should remain consistent across conversations while allowing natural response variation.[5]

Performance Under Load assesses how agent performance degrades under high-volume conditions, stress scenarios, or resource constraints, establishing operational limits and ensuring acceptable performance during peak usage. An e-commerce recommendation agent should maintain less than 2 second response times even when handling 10x normal traffic during sales events, while a customer support agent might experience graceful degradation from 1-second to 5-second responses under 5x load before requiring additional resources.[6] Translation agents should maintain accuracy above 85% even at 10x typical volume. Image generation agents use queue management when demand exceeds capacity to preserve quality without degradation.[7]

User Experience Metrics

For agents that interact directly with users, measuring satisfaction and experience quality provides essential insight into the practical effectiveness of the system. Research indicates that user experience metrics often diverge from technical performance measures, making them critical for comprehensive agent evaluation.[8]

Direct Feedback includes user ratings, surveys, and structured feedback mechanisms such as Customer Satisfaction (CSAT) scores and Net Promoter Scores (NPS). These metrics provide direct insight into user perceptions and satisfaction levels with agent interactions. For instance, a customer service chatbot might maintain a CSAT score of 4.2/5.0

while achieving 85% technical task completion, indicating strong user satisfaction despite some technical limitations.

Healthcare virtual assistants typically target NPS scores above 50, with leading implementations achieving scores of 60-70 through personalized interaction design.[9] Banking virtual assistants often track both immediate satisfaction (post-interaction rating) and longitudinal satisfaction (monthly surveys), with top-performing agents maintaining 90%+ positive ratings across both timeframes.[9]

Task Success from user perspective evaluates whether users feel their goals were accomplished through agent interactions, which may differ from technical task completion metrics. This user-centric view helps identify gaps between technical functionality and practical value delivery. A travel booking agent might technically complete 95% of reservation processes but achieve only 78% user-perceived success due to confusion about confirmation steps or unclear pricing communication. Educational tutoring agents demonstrate this disconnect well when they provide technically correct answers but fail to adapt explanations to student comprehension levels, resulting in 88% accuracy but only 65% student satisfaction with learning outcomes.

Engagement Metrics for conversational agents include conversation length, user retention rates, frequency of use, and interaction depth. These metrics indicate user acceptance and the practical utility of agent capabilities in real scenarios. High-performing customer support agents typically see average conversation lengths of 8-12 exchanges with 75% user retention for follow-up interactions, while mental health support agents might achieve deeper engagement with 15-25 exchange conversations and 85% weekly return rates.[10,11] E-commerce recommendation agents measure engagement through click-through rates, targeting

12-15% for personalized suggestions, and conversion rates, aiming for 8-12% improvement over non-personalized experiences.[12]

Qualitative Feedback involves analyzing user comments, reviews, support tickets, and other unstructured feedback to identify common pain points, feature requests, and improvement opportunities that quantitative metrics might not capture. Social media comments capture spontaneous, real-time opinions and emotions of customers, aiding in understanding immediate reactions and customer sentiment.

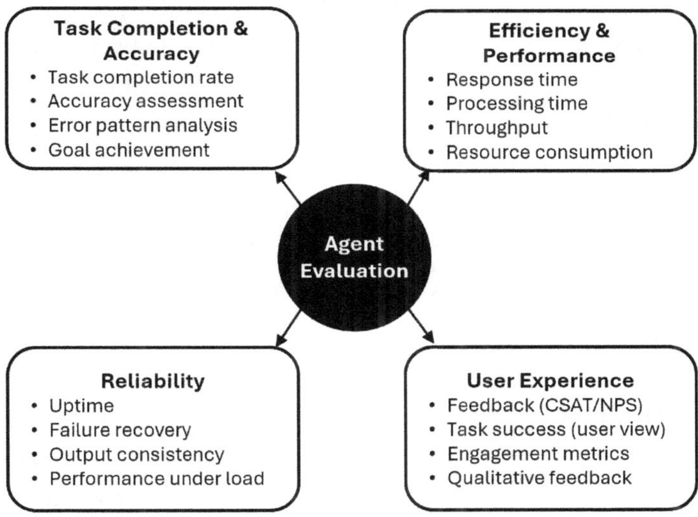

Figure 13: AI Agent Evaluation Framework

Considerations for Metric Selection

Effective metric selection requires a context-driven approach that chooses measurements directly relevant to the agent's specific purpose and business. Organizations must adopt a balanced measurement approach using combinations of metrics across different performance dimensions to avoid over-optimization in single areas that could compromise overall effectiveness—such as prioritizing speed at the expense of accuracy or user satisfaction. Finally, all selected metrics must be measurable and reliable, capable of consistent tracking over time to provide dependable indicators of performance changes.

Benchmarking Agent Effectiveness

Benchmarking transforms raw performance metrics into meaningful insights by comparing agent performance against relevant standards, providing essential context for identifying performance gaps and setting realistic improvement targets. Beyond simple comparison, benchmarking establishes data-driven goals based on achievable performance levels demonstrated by comparable systems, creating quantitative evidence that supports business cases and investment justification. However, effective benchmarking requires fair comparisons using similar datasets, workloads, and evaluation criteria to ensure meaningful and actionable insights.

Internal Benchmarking

Internal benchmarking provides the most directly relevant comparisons by evaluating agent performance against organizational baselines and historical data. This approach enables organizations to track progress systematically and identify optimization opportunities within their specific operational context.

Historical Performance tracks changes in an AI agent's performance over time by comparing current results against past performance to identify trends, detect performance drift, and analyze learning patterns. Document processing agents commonly exhibit a characteristic pattern where accuracy improves during initial deployment—for example, from 75% to 92% over the first six months as the system encounters and learns from edge cases—followed by stable performance that requires periodic retraining to maintain effectiveness. Exact performance gains vary widely across different implementations depending on factors like data quality, use case complexity, and system configuration.

Manual Process Comparison represents one of the most critical benchmarking exercises, comparing agent performance directly against the manual processes it replaced or augmented. This comparison typically encompasses speed, cost, accuracy, and quality dimensions, providing clear evidence of value delivery and return on investment. A financial services firm implementing an invoice processing agent might compare against baseline manual processing that required 15 minutes and costs $16 and error rate of 1.6% per invoice, versus less than 1 minute processing time, $3 cost by automated systems.[13]

A/B Testing enables systematic comparison of different agent versions, configurations, or approaches operating simultaneously under similar conditions. This might involve testing different underlying models, prompt strategies, workflow logic, or tool integrations to determine which approaches deliver superior performance on key metrics.

An e-commerce recommendation agent might test GPT-5 versus Claude-4 models over two weeks, discovering that GPT-5 generates 15% higher click-through rates, but Claude-4 produces 8% better conversion rates, leading to a hybrid approach optimized for different customer segments. Manufacturing quality control agents might undergo A/B testing between computer vision models, comparing YOLO-based detection achieving 91% defect identification versus ResNet-based approaches reaching 94% accuracy, with the statistical significance established through controlled testing environments.

External Benchmarking

External benchmarking provides critical market context and objective performance references by comparing agent capabilities against competitors, industry standards, and established best practices. This approach enables organizations to understand their relative market position and identify opportunities for competitive differentiation.

Competitive Benchmarking offers valuable insights into market positioning through competitor performance analysis when such data is available through public information, industry reports, or competitive analysis. Telecommunications companies often benchmark their technical support agents against competitor satisfaction scores published in industry surveys, discovering that while their agent achieves 4.1/5.0 customer ratings, market leaders consistently score 4.4/5.0, highlighting areas for improvement in conversation quality and problem resolution approaches.[14] E-commerce recommendation engines frequently analyze competitor conversion rate improvements, with market research indicating that leading implementations achieve 15-20% conversion rate improvements, providing performance targets for optimization efforts.[15]

Industry Standards evaluation assesses agent performance against industry-standards and widely adopted alternatives to determine whether performance meets market expectations. Healthcare diagnostic agents are benchmarked against FDA-approved clinical decision support systems that achieve 80-90% accuracy for specific conditions, establishing clear performance thresholds for regulatory compliance and clinical effectiveness.[16] Customer service implementations often compare against industry-standard platforms like Zendesk which report average case resolution times of 4.2 hours and first-contact resolution rates of 70-75%, establishing baseline performance expectations for competitive positioning.[17]

Academic research benchmarks leverage performance results from peer-reviewed research for similar agent types, often representing state-of-the-art performance levels that provide insight into theoretical and practical limits of current approaches. Computer vision agents benchmark against ImageNet classification accuracy where current state-of-the-art achieves 90.2% top-1 accuracy, or COCO object detection where leading models reach 58.2% mAP scores.[18,19]

Conversational AI systems compare against established benchmarks like SuperGLUE, where advanced models achieve scores as high as 89.3%, or against dialogue evaluation frameworks like ConvAI2 where top-performing systems demonstrate engagement scores of 3.1/4.0 and fluency ratings of 4.2/5.0.[20,21]

Continuous Improvement Strategies

Evaluation represents not a destination but a foundation for ongoing enhancement. Continuous improvements transform static systems into dynamic, evolving capabilities that adapt to changing requirements and deliver increasing value.

Iterative Development and Testing

Agile Development principles applied to agent improvement focus on continuous iteration through short development cycles. Teams regularly test agents against defined metrics and benchmarks, quickly identify performance gaps, and implement incremental improvements. This iterative approach allows organizations to adapt rapidly to changing business requirements while incorporating real-world operational insights into each development cycle, ensuring agents evolve continuously based on actual performance data and user feedback.

Systematic A/B Testing builds upon the testing approaches described previously by emphasizing rigorous analysis and structured evaluation protocols. These programs should include statistical analysis to ensure changes produce significant improvements, enabling evidence-based optimization decisions and preventing degradation of overall system performance.

Controlled Experimentation establishes formal processes for testing modifications in isolated environments before production deployment. These frameworks should include rollback procedures, performance monitoring, and clear criteria for promoting or rejecting changes based on measured impact.

User Feedback Integration

Comprehensive feedback collection gathers user input through multiple channels—surveys, ratings, comments, and support interactions. This qualitative data is systematically collected, categorized, and analyzed alongside quantitative metrics to provide a complete understanding of user experience and agent effectiveness.

Feedback-driven improvement prioritization uses both qualitative feedback and quantitative metrics to identify the most impactful improvement opportunities. Understanding user pain points helps prioritize development efforts toward changes that deliver the greatest user value and business impact. This approach focuses on user experience optimization by improving usability, functionality, and communication style based on user feedback patterns. Improvements may involve adjusting agent personality, improving response clarity, or adding capabilities that users frequently request.

Systematic Failure Analysis

Systematic failure analysis extends beyond simple error rate tracking through root cause analysis methodologies that deeply investigate specific failure instances to understand why agents fail to complete tasks, produce incorrect results, or receive negative feedback, which is essential for implementing effective corrections.

This comprehensive approach incorporates failure pattern recognition that identifies common failure modes, environmental conditions contributing to errors, and systemic issues affecting multiple interaction types, enabling proactive improvement efforts that address underlying causes, while systematic correction implementation ensures that insights from failure analysis translate into concrete improvements through model updates, prompt refinements, workflow optimizations, or additional training data incorporation that directly targets identified root causes.

Model and System Enhancements

Model Retraining and Fine-tuning programs ensure agents adapt to changing patterns through periodic updates with new data, reducing performance drift, and improving accuracy. A financial fraud detection agent might undergo monthly retraining incorporating new fraud patterns, improving detection rates from 91% to 95% while reducing false positives by 18%. Healthcare diagnostic agents typically update models every six months to incorporate new imaging techniques and diagnostic criteria, reaching up to 85% accuracy levels.[22]

Prompt Engineering Optimization focuses on continuously refining prompts that guide LLM behavior, as small changes in wording or structure can considerably impact performance. A legal document analysis agent improved contract clause extraction accuracy from 82% to 91% by leveraging prompt engineering techniques such as specific in-context examples and structured output formatting, as demonstrated in Scale AI's applications for contract data extraction. Technical support agents achieved a 30% reduction in escalation rates by introducing structured diagnostic prompts and systematic solution prioritization logic, processes similarly underpinned by best practices in data structuring and workflow optimization highlighted by Scale AI.[23]

Workflow and Logic Enhancement involves analyzing decision-making processes to identify bottlenecks and streamline operations. Banks have reduced customer onboarding processing times by reordering validation steps and implementing parallel document and identity verification, allowing the application process to be completed in as quickly as 10 minutes for individuals, while upholding strict compliance accuracy through digital and automated controls.[24]

Data Quality and Coverage Improvement ensures continuous enhancement of training and contextual data by adding relevant sources and refining cleaning processes. Retail inventory management agents have been shown to improve demand forecasting accuracy substantially—often by 10 to 15 percentage points—after incorporating diverse external data sources such as social media trends, weather patterns, and local event calendars. This integration allows the agent to capture a broader spectrum of demand signals and market dynamics, leading to more precise and adaptive inventory decisions.[25]

Proactive Monitoring

Proactive Monitoring systems establish comprehensive oversight through automated performance monitoring that tracks key metrics and triggers calibrated alerts when deviations occur, balancing sensitivity to significant issues while avoiding alert fatigue from minor fluctuations. These systems integrate trend analysis capabilities that leverage historical performance data to identify gradual changes indicating emerging issues or optimization opportunities, enabling proactive intervention before problems impact operations.

Continuous Monitoring integration embeds performance measurement directly into operational workflows. This integration provides immediate visibility into agent behavior and identification of issues requiring attention. The result is a holistic approach to maintaining optimal agent performance through data-driven oversight and predictive maintenance.

Systematic Improvement Cycle

The continuous improvement process follows a systematic cycle: Measure → Analyze → Improve → Measure. Each iteration should build upon previous learning, incorporate new insights from operational experience, and drive measurable performance enhancements. By regularly evaluating performance and acting on insights, organizations ensure their AI agents stay effective, adapt to new requirements, and become more valuable over time.

This systematic approach to continuous improvement transforms AI agents from static deployed systems into dynamic, learning capabilities that grow more valuable over time while maintaining alignment with changing business needs and user expectations.

9

Trust in Agents: Security, Safety, and Governance

"If people trust you, you're 90% there. If they don't, you're 100% nowhere."

— Howard Schultz

AI agents, with their autonomy and interaction with sensitive data and external systems, introduce a new frontier of vulnerabilities and ethical dilemmas, demanding a robust defense. Unlike the predictable, rule-based nature of conventional software, AI agents learn and make independent decisions, creating unique challenges that necessitate a comprehensive approach. To unleash the power of these intelligent systems, organizations must implement a comprehensive framework that includes strong security measures, commitment to safety and ethical principles, and governance for continuous oversight and control.

Protecting AI Agents

Security forms the technical foundation for safe and reliable AI agent deployment.

Threat Landscape

AI agents operate within complex ecosystems, creating multiple potential points of attack that require systematic identification and mitigation. The threat landscape encompasses both traditional cybersecurity risks and novel vulnerabilities specific to AI systems.

Data-Related Threats represent one of the most significant risk categories. Data breaches can occur when agents process or store sensitive information such as personal data, financial records, or proprietary business information. Security lapses leading to unauthorized access or data exfiltration can result in significant privacy violations, financial penalties, and brand damage.

These breaches may stem from vulnerabilities in data storage systems, insecure communication channels, or compromised access controls. Unauthorized access threats involve malicious actors attempting to gain control over the agent itself or the systems it interacts with, potentially allowing them to manipulate agent behavior, steal data, disrupt operations, or use the agent as a launching point for broader network attacks.

Model-Specific Attacks target the AI components that form the core intelligence of agents. Model poisoning attacks occur during the training phase, where attackers subtly manipulate training data to introduce hidden vulnerabilities or backdoors. A poisoned model may appear to function normally most of the time but behave incorrectly or maliciously when specific triggers are encountered, potentially leading to biased decisions, system failures, or security bypasses.

Adversarial attacks target AI models during their operational phase. Attackers craft carefully designed inputs that cause models to make incorrect predictions. These attacks take several forms. Evasion attacks modify inputs to cause misclassification. Extraction attacks involve repeatedly querying models to steal underlying algorithms or training data. Inference attacks aim to deduce sensitive information about training data. Prompt injection attacks are specific to LLM-based agents, where attackers manipulate input prompts to bypass safety instructions or execute unintended commands.

Infrastructure Vulnerabilities arise from the complex ecosystem in which agents operate. Insecure tool and API integrations present significant risks when agents rely on external services that may not be properly secured, potentially using hardcoded credentials or lacking adequate input validation. These integration points can become entry vectors for attackers to compromise either the agent or external systems. Denial of Service (DoS) attacks can overwhelm agents or their dependencies with excessive requests, rendering them unavailable for legitimate users and potentially causing cascading failures across interconnected systems.

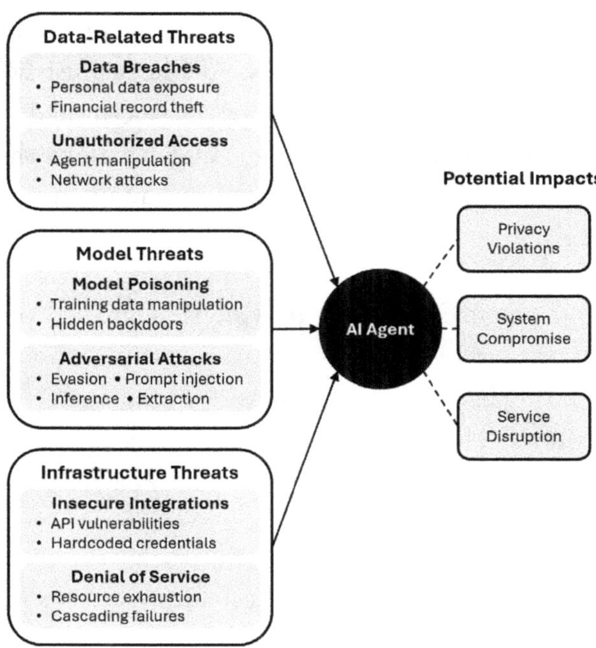

Figure 14: AI Agent Threat Landscape

Securing Agent Communication

AI agents rarely operate in isolation, communicating extensively with users, external APIs, databases, other agents, and various backend services. Securing these communication channels is vital for preventing eavesdropping, tampering, and impersonation attacks that could compromise the entire system's integrity.

Encryption and Secure Protocols form the foundation of communication security. All communication involving agents over public networks, must use strong encrypted protocols such as HTTPS with TLS encryption for web-based communication. Similar secure protocols should be implemented for other types of network connections to ensure that data exchanged between agents and other systems remains confidential and cannot be easily intercepted or modified in transit.

Authentication Mechanisms verify the identity of communicating parties and prevent unauthorized access. API keys must be kept confidential, rotated regularly, and configured with restricted permissions following the principle of least privilege. Organizations should avoid hardcoding keys directly in source code and instead use secure key management systems.

OAuth 2.0 and OpenID Connect provide standardized protocols for delegated authorization and authentication, useful for verifying user identity or granting agents limited access to user resources on external platforms.[1,2] Mutual TLS (mTLS) offers stronger authentication by requiring both the client agent and server to verify each other's identities using digital certificates.[3]

Access Control and Data Integrity mechanisms ensure that authenticated parties only access appropriate resources and that data remains unaltered during transmission. Authorization controls must enforce the principle of least privilege, ensuring agents and users only access specific resources and actions necessary for their functions through granular access controls. Data integrity checks using mechanisms such as digital signatures or message authentication codes (MAC) help ensure data has not been tampered with during transmission, while TLS protocols inherently provide some integrity protection.

Data Protection and Privacy

Beyond securing data in transit, protecting information when stored or processed by agents is equally critical given the sensitive data that many AI agents handle in their operations.

Data Storage Security requires comprehensive protection of all data repositories used by agents. Encryption at rest ensures that sensitive data stored in databases, files, and logs remains protected using strong encryption algorithms, making data unreadable even if unauthorized access to storage media occurs. These security measures are essential for meeting regulatory requirements under frameworks like GDPR, which mandate appropriate technical and organizational measures to protect personal data.

Effective key management practices are essential for maintaining the security of encryption systems. Secure storage practices involve utilizing secure database configurations, hardened file systems, and cloud storage services with robust security features, while implementing strict access controls on all storage systems.

Access Control and Data Governance establish frameworks for managing data access and handling throughout the agent lifecycle. Access control systems must ensure that only authorized personnel and processes can access agent data stores and configurations, implementing role-based access control (RBAC) systems with regular permission reviews.

Data minimization principles require collecting and retaining only essential data for agent functionality, aligning with privacy principles, and avoiding unnecessary storage of sensitive information. This approach directly supports GDPR's and CCPA's data minimization requirements. Where possible, organizations should implement anonymization or pseudonymization techniques to reduce privacy risks associated with sensitive data processing.

Data Handling Policies and Validation provide frameworks for secure data management. These policies establish clear procedures for collecting, processing, storing, and deleting sensitive data. They also include secure disposal practices, retention guidelines, and personnel training requirements. These policies must support individual rights under GDPR and CCPA, such as the right to be forgotten or deleted, ensuring agents can facilitate data subject requests and meet regulatory timelines. Input and output validation and sanitization processes help prevent injection attacks and data leakage by rigorously checking and cleaning all data inputs received by agents and outputs generated by them.

Defense Against Adversarial Attacks

AI models are the brains of many agents, making model protection against adversarial attacks a critical security concern that requires specialized approaches.

Defense Strategies require layered approaches since no single defense mechanism provides complete protection. Input protection mechanisms include rigorous input validation and sanitization to check and clean inputs before feeding them to models, with specific attention to detecting and neutralizing potential prompt injection attempts in LLM-based systems. Implement these protections by establishing input parsing pipelines that scan for malicious patterns, suspicious command sequences, and injection attempts before data reaches the model.

Deploy automated filters that recognize common attack vectors such as role-playing prompts, system message overrides, and delimiter-based injections. Rate limiting and input quotas help mitigate model extraction attacks by restricting the number of queries external users can make to agents. Configure these limits based on user authentication levels, implementing sliding window rate limiters and CAPTCHA challenges for suspicious activity patterns.

Model Hardening techniques strengthen the underlying AI models against adversarial manipulation. Adversarial training involves including adversarial examples in training data to make models more robust against attacks during inference. Implement this training by generating adversarial examples using techniques like FGSM (Fast Gradient Sign Method) or PGD (Projected Gradient Descent), then incorporating 10-20% adversarial samples into training batches.

Model hardening techniques such as defensive distillation or feature squeezing reduce model sensitivity to small input perturbations. Apply defensive distillation by training a second model to match the softmax outputs of the original model at higher temperatures, then deploying the distilled version. Implement feature squeezing through bit-depth reduction, spatial smoothing, or JPEG compression to remove adversarial perturbations.

Differential privacy adds controlled noise during training or inference to protect individual data point privacy in training sets, making inference attacks more difficult. Configure differential privacy by adding calibrated Gaussian noise to gradients during training or to model outputs during inference, with epsilon values typically between 0.1 and 10.

Output Protection mechanisms are crucial for monitoring and filtering the responses of AI agents, notably those with generative capabilities. Implement multi-layered content filtering by deploying real-time classifiers that scan outputs for harmful content categories including violence, hate speech, personal information, and inappropriate instructions. Use both rule-based filters such as regex patterns and keyword blacklists, and ML-based classifiers trained on harmful content datasets. Deploy semantic analysis tools that evaluate output coherence and detect potentially manipulated responses that deviate from expected behavioral patterns.

Establish output validation pipelines that check responses against predefined safety criteria before delivery to users. Implement confidence thresholding where outputs below certain confidence scores trigger human review or safer fallback responses. Deploy anomaly detection systems that continuously monitor output patterns, flagging unusual response lengths, sentiment changes, or topic deviations that could indicate ongoing attacks.

Set up real-time monitoring dashboards that track output quality metrics, user feedback, and automated safety scores. Configure automated response mechanisms that can temporarily disable agents, switch to more conservative models, or route to human oversight when suspicious outputs are detected. Maintain comprehensive logging of all inputs, outputs, and filtering decisions to enable post-incident analysis and continuous improvement of protection mechanisms.

Continuous Security Monitoring

Security is an ongoing process requiring continuous vigilance, monitoring, and adjustment to evolving threats and changing operational requirements.

Monitoring and Logging systems provide the foundation for security awareness and incident response. Comprehensive logging must capture all significant agent actions, decisions, API calls, errors, and security-relevant events, creating essential documentation for monitoring, debugging, and forensic analysis following security incidents. Logs themselves must be stored securely to prevent tampering or unauthorized access. Real-time monitoring systems track agent behavior, performance, resource usage, and network traffic continuously, looking for anomalies, unexpected activity spikes, high error rates, or patterns indicative of security threats such as repeated failed login attempts or unusual API call sequences.

Security Assessment and Testing provides proactive identification of vulnerabilities and weaknesses. Regular security audits review agent architecture, code, configurations, access controls, and adherence to security policies. Vulnerability scanning and penetration testing exercises regularly scan agent infrastructure and code for known vulnerabilities while conducting controlled attacks to identify security weaknesses proactively.

Maintenance and Incident Responses ensure ongoing security posture and effective response to security events. Dependency management requires keeping all underlying software, libraries, and frameworks, including agent frameworks and operating systems, updated with the latest security patches. Comprehensive incident response plans detail specific steps to take during security breaches or suspected attacks, including containment, eradication, recovery, and post-incident analysis procedures.

Security Domain	Control	Implementation
Communication	Encryption	HTTPS/TLS, secure protocols
	Authentication	API keys with restrictions/rotation, OAuth 2.0, mTLS
	Access Control	Least privilege, digital signatures, MAC, TLS integrity protection
Data Protection	Storage Security	Encryption at rest, secure databases, hardened file system, cloud with security
	Data Governance	RBAC systems, data minimization, anonymization
	Policy & Compliance	GDPR/CCPA procedures, input validation
Adversarial Defense	Input Protection	Parsing pipelines, rate limiting, CAPTCHA
	Model Hardening	Adversarial training with FGSM/PGD, adversarial sampling, defensive distillation
	Output Protection	Content filtering, semantic analysis tools, anomaly detection, logging
Continuous Monitoring	Monitoring & Logging	Comprehensive action/decision logs, resource use, network traffic
	Assessment & Testing	Security audits, penetration testing
	Maintenance & Response	Dependency updates, incident response

Table 9: Security Controls and Implementation Methods

Safety and Ethics

While security protects against external threats, safety and ethics address the fundamental responsibility of ensuring AI agents operate without causing harm through their intended functionality.

Identifying Risks

The deployment of AI agents introduces unique risks, requiring careful consideration and comprehensive mitigation strategies.

Bias and Discrimination represent some of the most significant challenges in the deployment of AI agents. AI agents trained on datasets reflecting historical societal biases can inadvertently perpetuate or amplify unfair discrimination. For example, an agent used for resume screening might learn to favor candidates from certain demographics, while a loan application agent might unfairly deny applicants based on proxies for protected characteristics learned from biased training data. These risks undermine fairness principles and can have significant consequences for individuals and groups, potentially reinforcing or exacerbating existing societal inequalities.

To identify bias, conduct bias audits of training data and agent outputs across demographic groups. Use fairness metrics like demographic parity and equalized odds. Perform A/B testing with diverse scenarios and engage stakeholders to review agent behavior patterns.

Privacy and Data Protection risks arise from agents' extensive data requirements for effective operation. Agents need access to large amounts of data to work effectively, often sensitive personal information. Healthcare agents access patient records, while financial agents process transaction histories. Risks include data mishandling, exposure through security breaches, or usage in ways users did not consent to, leading to significant privacy violations. Additionally, agents' ability to infer new information from combined data sources can create unforeseen privacy risks that may not be apparent from individual data sources alone.

Conduct data mapping and privacy impact assessments (PIAs) to identify privacy and data protection risks. Perform inference risk analysis to identify what sensitive information could be derived from data combinations. Review data access logs, retention practices, and consent mechanisms. Assess security vulnerabilities and regulatory compliance gaps.

Operational Risks stem from the complex and sometimes unpredictable nature of AI systems. The behavior of complex agents, such as those involving deep learning or intricate reasoning loops, can sometimes be difficult to predict or fully understand, even for their creators, potentially leading to unexpected and harmful actions in novel situations. This edge case problem is compounded by accountability gaps that arise when autonomous agents cause harm, making it challenging to determine responsibility among developers, deployers, users, data providers, or the agents themselves who lack legal personhood.

To find the operational risks, implement comprehensive testing including edge cases, stress testing, and adversarial scenarios. Use model interpretability tools and failure mode and effects analysis (FMEA). Monitor performance metrics and behavioral patterns for anomalies. Establish clear accountability frameworks and conduct regular operational audits.

Societal Risks refer to the far-reaching impacts that can arise when intelligent agents are deployed at scale—affecting institutions, public trust, employment, and the fabric of daily life. While agents can augment human capabilities, widespread automation of cognitive tasks previously performed by humans raises concerns about job displacement and the need for workforce reskilling. Additionally, agents designed for beneficial purposes can potentially be repurposed for harmful ones, such as using autonomous agents capable of online interaction to spread misinformation at scale, manipulate social opinions, or execute coordinated cyberattacks.

Here is how you identify these risks. Conduct impact assessments on job displacement potential in collaboration with workforce development experts. Perform misuse case analysis to identify harmful repurposing possibilities. Monitor public discourse, conduct scenario planning, and collaborate with policymakers and ethicists to understand broader implications.

Ethical Principles

Ethical AI agent development must be anchored in foundational principles that prioritize human benefit, safeguard individual rights, and uphold societal values.

Core Ethical Principles provide the foundation for responsible AI development. The principle of beneficence and non-maleficence requires that agents be designed to do good while avoiding harm, whether physical, psychological, financial, or social. This involves anticipating potential negative consequences and building appropriate safeguards into agent design and operation.

Agents must treat all individuals and groups equitably, avoiding bias through careful data selection, algorithm design, and impact assessments. Privacy principles require that agents respect individual privacy by handling personal data securely, transparently, and only for legitimate purposes, adhering to data minimization principles and obtaining appropriate consent.

Accountability Principles ensure clear responsibility and oversight frameworks. Organizations must establish defined lines of responsibility for agent actions and outcomes, with accessible mechanisms for redress when harm occurs.

Human-Centric Principles maintain the central role of human agency and oversight in AI systems. Human autonomy principles require that agents augment human decision-making and control, most critically in high-stakes situations, with users retaining meaningful control and the ability to override agent actions when appropriate. Human oversight ensures appropriate supervision to monitor agent behavior, intervene when necessary, and maintain alignment with goals and values.

Transparency and Explainability

As AI agents make increasingly complex decisions, understanding how they made those decisions becomes critical for building trust, enabling debugging, ensuring accountability, and meeting regulatory compliance requirements.

Understanding Transparency encompasses multiple dimensions of system openness and clarity. Data transparency involves knowing what data agents were trained on and what data they access during operation, providing stakeholders with understanding of potential biases and limitations. Algorithmic transparency enables understanding the algorithms and models used within agents, including their capabilities, limitations, and potential failure modes. Process transparency provides visibility into operational workflows of agents, including which tools they use, what intermediate steps they take, and how they reason through problems, often facilitated through detailed logging and tracing systems.

Explainability provides clear, human-friendly explanations for how AI agents make decisions and produce outputs. It builds trust by helping users understand why agents made specific recommendations, making them more likely to adopt suggestions. Also, it enables developers to debug problems and improve performance by revealing why agents failed. Explainability also supports accountability and audit processes when issues arise, while ensuring compliance with regulations requiring explanation of automated decisions. Finally, explainability promotes fairness by exposing potential biases in AI decision-making, allowing teams to identify and address unfair patterns.

A practical example of explainability comes from my experience with an automatic suture inspection system at a Fortune 50 pharmaceutical company, implemented by EazyML. The AI system provided confidence intervals alongside inspection results, automatically passing outcomes with 95% or higher confidence levels. This approach built trust in the quality inspection team by clearly communicating system certainty and achieved error rates below 1%. The confidence scoring helped operators understand when human review was needed.

Balancing explainability with system performance is an important consideration for agents. Achieving full transparency and explainability can be difficult with complex models like deep neural networks or LLMs that often function as black boxes. The reasoning processes of agents involving multiple steps and tool calls can also be complex to trace, creating trade-offs between model performance and explainability. Organizations can address these challenges through intrinsic explainability, post-hoc explanation techniques, detailed logging and tracing, and clear documentation of agent design, data sources, and intended logic.

Governance of AI Agents

As organizations deploy increasing numbers of agents making decisions and taking actions across their operations, robust governance mechanisms become critical for ensuring responsible operation, and effective oversight at scale.

Establishing Consistent Standards

Effective governance requires centralized repositories for all governance rules—ethical guidelines, safety protocols, security standards, data policies, and compliance requirements. This ensures consistency across deployments and provides a single source of truth for updates. An AI Center of Excellence (CoE) should serve as the authoritative body for policy development and enforcement, combining technical, legal, ethical, and business expertise.

Centralized policy management systems must support version control, policy templates, and integration with development pipelines for automated compliance evaluation. The framework should include clear conflict resolution mechanisms and escalation procedures for policy ambiguities. Regular review processes ensure governance evolves with technology, regulations, and organizational needs.

Compliance and Auditing

Organizations must integrate automated compliance validation into CI/CD pipelines, evaluating agent configurations, prompts, permissions, and behavioral constraints against governance frameworks before production deployment. Runtime monitoring systems flag policy violations, enabling real-time enforcement and immediate intervention.

Comprehensive audit trails provide accountability through detailed, immutable logs of agent actions, decisions, data access, and tool usage. These logs demonstrate regulatory compliance, enable post-incident analysis, and identify behavioral patterns for improvement.

Fairness and accountability require systematic bias identification and mitigation across training data, algorithms, and feedback loops. Organizations must employ fairness metrics and auditing tools to test for demographic biases, implementing mitigation strategies including data preprocessing, algorithm modifications, output adjustments, and diverse development teams.

Regular auditing validates policy compliance, guardrail effectiveness, and audit trail accuracy while assessing alignment with intended purposes and identifying behavioral drift. Audits also refine governance frameworks and improve oversight mechanisms.

Accountability frameworks establish clear organizational roles and responsibilities across the agent lifecycle—design, development, testing, deployment, operation, and oversight. These frameworks include policies for ethical review, risk assessment, ongoing monitoring, and comprehensive logging systems to enable investigation when issues arise.

Access Controls

Organizations must implement granular access controls defining what systems, data, and actions each agent can access, following least-privilege principles that limit permissions to necessary functions only. Access controls include specifying external system interactions and autonomy levels across different operational contexts.

Access control systems manage agents throughout their lifecycle—provisioning, operation, and decommissioning. Regular reviews ensure permissions remain appropriate as agents and requirements evolve. Automated permission management maintains consistency while reducing administrative burden across large agent deployments.

Human access controls require appropriate separation of duties and accountability for those who design, deploy, manage, or interact with agents. Clear roles for developers, data scientists, operations personnel, and business users must include defined access rights and oversight responsibilities. Emergency access procedures enable human intervention while maintaining proper audit trails and approval processes.

Continuous Learning

Agent ecosystems require continuous learning through systematic feedback loops that capture insights, analyze patterns, and implement improvements to enhance effectiveness over time.

Feedback collection encompasses multiple sources. User feedback includes ratings, comments, success rates, and corrections. Operational feedback covers performance metrics, error logs, and resource consumption. Business feedback tracks KPIs and ROI metrics to assess strategic alignment and organizational impact.

Analysis and routing transforms raw data into actionable intelligence by categorizing and routing feedback to appropriate teams—developers, prompt engineers, center of excellence, or business stakeholders. Analytics identify trends, systemic issues, and common failure points across the agent fleet.

Implementation translates insights into concrete improvements: updating prompts and templates, refining agent logic and orchestration, improving tool reliability, retraining models, and updating governance policies. Organizations must prioritize improvements based on impact and implementation effort to maximize value while managing resources effectively.

This continuous cycle ensures agent ecosystems evolve dynamically, maintain alignment with changing business needs and user expectations, and uphold safety, security, and ethical standards.

Akka: Enabling Security, Safety, and Governance

Enterprise-Grade Security and Reliability

Akka's distributed actor model enables agents to run as lightweight, isolated processes communicating over brokerless, encrypted channels. Built-in supervision strategies and durable persistence ensure workflows recover from failures without data loss. Integration with Prometheus and OpenTelemetry provides real-time monitoring and detailed audit logs, supporting strong security awareness and incident response.

Safety through Structured Orchestration

Akka supports complex human-in-the-loop workflows and multi-agent evaluation patterns. Agents can critique or verify each other's outputs, embedding checks and balances into automated decision-making. This design reduces errors and supports safer, more predictable behavior at scale, even in mission-critical environments.

Governance and Compliance Controls

Akka includes audit trails, versioned rolling upgrades, and flexible deployment options across cloud or on-premises environments. Supervision hierarchies enforce approved, tested behaviors in production, while centralized logging and observability enable traceability and regulatory compliance. These features help organizations maintain strict governance standards for AI operations.

Figure 15: Secure and Governable AI with Akka

Regulatory Landscape

The rapid advancement of AI has spurred governments and regulatory bodies worldwide to develop frameworks governing AI use, creating a complex and evolving landscape that organizations must navigate carefully to ensure compliance and avoid legal risks.

Data privacy and protection regulations form the cornerstone of contemporary AI governance, establishing essential compliance obligations for data-intensive AI agents. Foundational laws such as the EU's General Data Protection Regulation (GDPR) and the California Consumer Privacy Act (CCPA) impose stringent requirements on how personal data is collected, processed, and stored—directly shaping the design and operation of AI systems.[4,5] These regulations establish four key requirements: explicit user consent for data collection and AI processing, data minimization that limits collection to necessary information only, robust security measures throughout the data lifecycle, and guaranteed user rights including access, correction, portability, and deletion. To ensure compliance across multiple jurisdictions, organizations often adopt the most restrictive standards as a baseline, enabling consistent governance of cross-border data flows and local regulatory nuances. For operationalizing these legal mandates, the DAMA-DMBOK (Data Management Body of Knowledge) offers a comprehensive framework that supports data quality, privacy management, and lifecycle governance. By translating abstract legal requirements into structured data practices, DAMA-DMBOK enables organizations to build responsible AI agent ecosystems that are both compliant and resilient.[6]

The EU AI Act proposes a comprehensive risk-based approach, categorizing AI systems based on their potential risk levels—unacceptable, high, limited, minimal—and imposing corresponding requirements for development, transparency, and oversight for high-risk applications.[7] Similar regulatory frameworks are being developed or proposed in other jurisdictions, though approaches vary considerably in scope and implementation requirements.

Technical standards and frameworks provide additional guidance for AI governance. The NIST AI Risk Management Framework offers a comprehensive approach to identifying, assessing, and managing AI risks throughout the system lifecycle.[8] OECD AI Principles establish international guidelines for trustworthy AI, emphasizing human-centered values, transparency, and accountability.[9] ISO/IEC 23053 provides guidance on AI risk management,[10] while ISO/IEC 23894 addresses AI risk management for organizations.[11]

Sector-specific regulations and disclosure requirements introduce added layers of compliance obligations. Industries such as finance and healthcare often have specific regulations governing algorithmic decision-making, data privacy, and system reliability that apply to AI agents operating within those domains. Emerging regulations frequently include transparency requirements, such as disclosing when users are interacting with AI systems or providing explanations for automated decisions. Existing anti-discrimination laws apply to AI agent decisions, requiring organizations to ensure their systems comply with equal treatment requirements across protected categories.

Organizations must stay informed about relevant legal and regulatory requirements in all jurisdictions where they operate, as compliance is not optional. Effective compliance involves understanding applicable laws, implementing necessary technical and organizational measures, and often requires ongoing legal consultation to navigate the evolving regulatory landscape effectively.

Successful AI agent deployment requires integrated security, ethics, and governance, supported by diverse expertise and ongoing evaluation rooted in human and societal values. As agents grow more autonomous, these frameworks become essential to unlocking AI's potential while safeguarding trust and minimizing risk.

10

Leadership Journey: A Roadmap for Implementation

"The journey of a thousand miles begins with a single step."

— Lao Tzu

We have explored AI agents through their architecture, practical applications, ROI, and safety. The challenge now is converting that understanding into meaningful business impact. This chapter presents a practical roadmap to guide organizations from concept to deployment—bridging theory and execution to deliver tangible outcomes.

Aligning Technology with Purpose

Embarking on AI agent development without strategic planning risks misalignment and wasted investment. A methodical approach ensures focus, alignment with business goals, and measurable value from implementation.

Organizational Alignment

The foundation of successful AI agent initiatives rests on establishing clear connections between proposed agent capabilities and overarching organizational objectives. This alignment process requires systematic evaluation of how AI agents can directly support strategic priorities, such as improving customer satisfaction, increasing operational efficiency, or accelerating innovation cycles.

Effective alignment begins with comprehensive stakeholder engagement to understand current organizational challenges, strategic priorities, and success metrics. Leadership teams must articulate how AI agent capabilities will contribute to achieving specific business outcomes, ensuring that proposed initiatives receive appropriate support and resource allocation.

This alignment enables organizations to develop compelling business cases and establish formal evaluation processes that prioritize AI agent initiatives with the greatest potential for ROI, strategic impact, and organizational value creation. This approach prevents the common pitfall of pursuing technically interesting but strategically irrelevant agent implementations.

Use Case Prioritization

Not every organizational task or process represents an appropriate candidate for AI agent implementation. Successful adoption requires systematic identification and evaluation of potential applications across different departments, functions, and operational areas using comprehensive assessment criteria that balance potential value against implementation feasibility.

Value assessment methodology evaluates the potential business impact of proposed AI agent implementations across multiple dimensions. Financial impact considerations include potential cost savings through automation, revenue generation opportunities through new products or services, and productivity improvements that enable resource reallocation to higher-value activities. Strategic impact evaluation examines how agent capabilities might provide competitive advantages, enable new business models, or enhance organizational capabilities that support long-term strategies.

Feasibility analysis framework assesses the practical implementation requirements and constraints for each proposed use case. Technology maturity evaluation determines whether required AI capabilities are sufficiently developed and reliable for production deployment. Data availability and quality assessment examines whether necessary training and operational data exists, is accessible, and meets quality standards required for effective agent performance. Organizational readiness analysis evaluates whether the organization possesses the technical expertise, infrastructure, and change management capabilities necessary for successful implementation.

Complexity and risk assessment examines the technical, operational, and organizational challenges associated with each potential use case. Reference to the complexity management principles outlined in previous chapters helps organizations understand the full scope of implementation requirements and potential pitfalls. This assessment should include evaluation of integration requirements, data dependencies, performance expectations, and potential failure modes that could impact successful deployment.

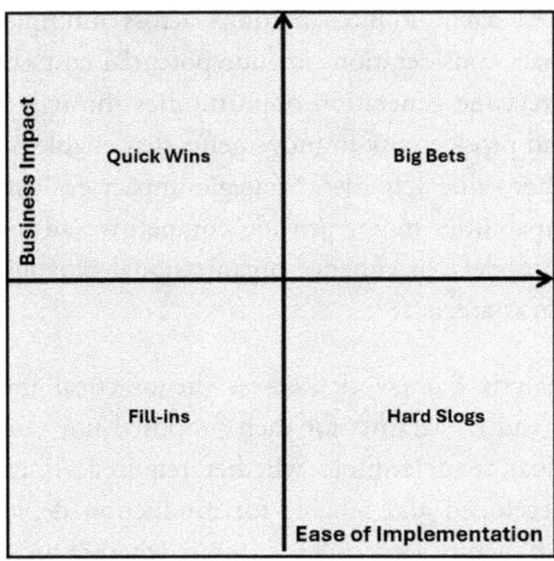

Figure 16: Use Case Prioritization

Initial project selection should prioritize opportunities that offer optimal balance between high potential value and manageable implementation complexity, creating early wins that build organizational confidence and expertise while establishing foundations for more ambitious future initiatives.

For detailed steps in choosing the use cases and creating success, please read *How to Choose the Right AI Use Cases.*[1]

Success Criteria

Each selected use case requires specific, measurable objectives that provide clear guidance for development efforts and enable objective evaluation of implementation success. The SMART framework guides AI agent objectives to be Specific, Measurable, Achievable, Relevant, and

Time-bound, while also setting clear performance benchmarks to evaluate success.

Effective objectives should encompass three measurement categories: technical capabilities such as accuracy rates and response times, operational measures such as throughput and reliability, and business impact indicators such as cost reduction and user satisfaction. For example, "achieve 95% accuracy in automated document processing tasks while maintaining processing speed of 100 documents per hour" establishes both technical and operational benchmarks.

Each objective must define three performance thresholds: minimum acceptable performance that meets basic requirements, target performance levels that indicate successful implementation, and stretch goals that represent exceptional success. These metrics and thresholds should be documented formally and agreed upon by all stakeholders before development begins to ensure alignment and prevent scope creep during implementation. This comprehensive framework enables objective assessment of agent performance while providing clear guidance for development priorities and resource allocation.

Implementation Planning

Successful AI agent implementation requires comprehensive planning that addresses development timelines, resource requirements, risk mitigation strategies, and deployment approaches that ensure sustainable operation and continuous improvement.

Timeline development involves creating realistic schedules for development, testing, and deployment phases that account for iterative development, stakeholder feedback and potential challenges. Agile development methodologies often prove most effective for AI agent projects, enabling adaptive responses to changing requirements and rapid incorporation of learning from testing and user feedback.

Resource allocation encompasses budget planning for tools, APIs, cloud services, and personnel. Organizations must plan for human resources including developers, data scientists, domain experts, and project managers, as well as infrastructure requirements for development, testing, and production environments. Additionally, ongoing operational costs including maintenance, monitoring, and continuous improvement efforts require careful consideration.

Risk assessment and mitigation proactively identifies potential challenges and establishes strategies for addressing them before they impact project success. This assessment should encompass technical risks such as model performance limitations, operational risks such as integration challenges, and organizational risks such as user adoption barriers. The comprehensive risk management approaches detailed in previous chapters regarding ethics, security, and safety provide essential frameworks for this assessment.

Strategic Sourcing Framework

Organizations face a fundamental strategic decision regarding how to acquire AI agent development capabilities, with three primary acquisition paths offering distinct advantages and challenges. The choice here influences investment timelines, resource requirements, competitive advantage potential, and long-term organizational capabilities.

Build

This approach leverages foundational models and open-source frameworks to deliver tailored solutions tailored that meet specific organizational needs.

Strategic advantages of in-house development include complete control over agent design, functionality, and data handling practices, enabling deep customization to unique business requirements and processes. Organizations develop valuable internal expertise and intellectual property that can provide sustainable competitive advantages while maintaining maximum flexibility to adapt and evolve agents as business requirements change over time.

Implementation requirements for successful in-house development include acquiring specialized AI/ML expertise that may be scarce and expensive. Organizations must also commit to longer development timelines compared to alternative approaches and assume full responsibility for ongoing maintenance, security updates, and technical evolution of agent capabilities.

The in-house approach makes most strategic sense for organizations with unique, complex requirements that cannot be effectively addressed by existing solutions, those operating in highly regulated industries requiring complete control over data and algorithms, or companies where AI agent capabilities represent a core competitive differentiator. Additionally, organizations with existing technical infrastructure and AI expertise, substantial long-term investment capacity, and business models that can absorb extended development timelines are well-positioned to benefit from this approach.

Buy

With the buy approach purchase or subscribe to pre-built AI agent solutions and platforms from established vendors who have already developed solutions for common business use cases.

Strategic benefits include faster deployment timelines and quicker time-to-value realization, often with lower initial capital requirements compared to building custom solutions. Organizations gain immediate access to vendor expertise, ongoing support, and maintenance while leveraging solutions specifically designed and optimized for common use cases such as customer service automation or sales forecasting.

Strategic limitations encompass reduced customization options that may not perfectly align with unique organizational requirements, potential vendor lock-in situations that limit future flexibility, and ongoing subscription costs that accumulate over time. Organizations also have limited control over data handling practices and underlying agent logic while facing potential integration challenges with existing systems and processes.

The commercial solution approach is ideal for organizations with standardized use cases, limited AI development resources, or urgent deployment needs. It offers a fast, low-risk path to value—especially for businesses in competitive markets, with constrained budgets, or those prioritizing predictable costs and proven vendor support.

Augment

The augment strategy leverages AI components such as LLM APIs and specialized ML libraries, combined with integration frameworks like Akka or LangChain, to enhance existing systems or build targeted intelligent features —without the need to create entirely new standalone agent capabilities.

Strategic value proposition includes providing a more incremental and less disruptive path to AI adoption while leveraging existing infrastructure investments, established data repositories, and proven business processes. Organizations can achieve focused improvements on specific operational pain points while often realizing more cost-effective results than comprehensive build or buy strategies for targeted functionalities.

Implementation considerations still require in-house or contracted development expertise for successful execution, and integration complexity can be significant depending on existing system architecture and data accessibility. The augment approach may not achieve the full transformative potential of purpose-built agents but provides excellent opportunities for incremental capability enhancement and organizational learning.

The augment approach is especially strategic for organizations with robust existing systems that benefit from intelligent enhancements without requiring full replacement. It suits teams aiming to minimize disruption while exploring AI capabilities and is well aligned with companies that have moderate technical resources and established development teams.

Augmentation strategy benefits organizations pursuing gradual AI adoption to build internal expertise and confidence, those with complex legacy systems where full replacement is impractical, or businesses wanting quick wins on specific pain points while maintaining operational continuity. Additionally, organizations with limited AI budgets, risk-averse cultures, or needing to demonstrate AI value before larger investments will find the augment approach ideally suited to their requirements.

The optimal sourcing strategy depends on multiple organizational factors including the complexity of use cases and implementation, internal technical expertise, budget constraints, competitive pressures, regulatory requirements, and the availability of suitable commercial solutions.

Building Internal Expertise

Successful AI agent adoption depends on developing and maintaining appropriate expertise within the organization. Technology alone cannot drive successful implementation. Organizations must systematically cultivate the knowledge, skills, and capabilities necessary to design, develop, deploy, and operate AI agents effectively over time.

AI Literacy

Building widespread AI literacy across key areas of an organization lays the groundwork for successful agent adoption. It helps stakeholders grasp what AI can and can't do—and understand how it affects their specific roles and responsibilities.

Organizational education programs should foster a foundational understanding of AI concepts and practical limitations among both technical and non-technical staff. This foundational knowledge fosters better collaboration between different organizational functions and helps identify additional suitable use cases that might not be apparent to purely technical teams.

Cross-functional understanding enables better integration between AI agent capabilities and existing business processes by ensuring that domain experts understand how agents can enhance their work while technical teams understand the business context and requirements that drive successful implementation.

Technical Skills

AI agent development requires specific technical competencies that must be systematically developed within the organization or acquired through strategic hiring and partnership approaches.

Core programming competencies include proficiency in languages such as Python, Java or JavaScript that form the foundation for most AI agent development frameworks and tools. These programming skills enable teams to customize agent behavior, integrate with existing systems, and implement necessary monitoring and maintenance capabilities.

Fundamental AI/ML knowledge includes core concepts in machine learning, model types, training methods, and evaluation techniques—all essential for designing and optimizing effective agents. This knowledge enables teams to make informed choices about model selection, performance optimization, and troubleshooting approaches.

Agent framework expertise involves developing proficiency with specialized tools and platforms such as Akka, LangChain or similar frameworks that accelerate agent development while providing proven patterns and capabilities for common agent functionalities.

Prompt engineering and LLM interaction skills represent critical capabilities for organizations leveraging large language models in their agent implementations. These skills enable effective communication with AI models and optimization of agent responses for specific use cases and user requirements.

Data science skills in data collection, cleaning, preparation, and analysis ensure that agents have access to high-quality information necessary for effective operation. Data skills also support ongoing performance monitoring and optimization efforts.

Integration skills enable teams to connect agents with existing systems, implement robust software architectures, and ensure reliable operation in production environments.

Explore a curated list of free or low-cost resources to build your AI literacy skills at reddymallidi.com.

Organizational Learning

Structured training programs including workshops, certifications, and formal education opportunities help existing employees develop necessary skills while demonstrating organizational commitment to AI capability development. These programs should combine theoretical knowledge with practical application opportunities.

Cross-functional team formation brings together technical experts, domain specialists, data professionals, project managers, and representatives from legal, ethics, and security functions to ensure comprehensive consideration of all implementation aspects. This collaborative approach leverages diverse perspectives while building shared understanding across organizational boundaries.

Hands-on learning approaches encourage skill development through practical application in pilot projects and controlled experiments that allow teams to gain experience with tools and methodologies while contributing to organizational objectives. These learning opportunities build confidence and expertise while delivering tangible business value.

Experimentation culture creates organizational environments where teams feel empowered to explore new approaches, learn from failures, and share knowledge about effective practices and potential pitfalls. This culture supports continuous learning and growth as AI technologies continue to evolve rapidly.

Establishing AI COE

As AI agent initiatives expand across organizational functions, establishing a dedicated Center of Excellence (CoE) provides essential structure, coordination, and governance capabilities that ensure consistent quality, strategic alignment, and effective resource utilization across multiple projects and departments.

Purpose and Value Proposition

An AI Agent Center of Excellence (CoE) serves four critical functions that help organizations maximize the value of their AI investments. These functions ensure that AI initiatives are scalable, responsible, and aligned with strategic goals.

Best practice development consolidates organizational learning by capturing successful implementation patterns, reusable code libraries, proven prompt templates, and standardized development practices. This foundation accelerates future projects, reduces duplication of effort, and helps teams avoid repeating past mistakes.

Quality and consistency implementation ensures that agents meet established standards for performance, security, ethics, and user experience. This is achieved through consistent methodologies and robust governance processes that promote reliability and trust across deployments.

Governance and resource optimization play a vital role in sustaining long-term success. Governance involves creating and enforcing policies for ethical AI use, regulatory compliance, data protocols, and model validation. Meanwhile, resource optimization provides visibility into ongoing projects, facilitates cross-team collaboration, prevents duplication, and ensures alignment with broader organizational priorities.

Organizational Structure

The CoE requires four distinct functional areas to operate effectively. The leadership and strategic oversight function is handled by AI strategists and program managers, who align agent initiatives with organizational objectives, manage resource allocation, and ensure strategic coherence across projects.

Technical expertise and platform management encompasses lead AI/ML engineers, data scientists, and platform architects who establish technical standards, manage shared infrastructure, evaluate new technologies, and provide expert consultation to project teams.

Governance and risk management involves ethicists, security specialists, legal experts, and compliance professionals who ensure agent implementations meet standards for responsible AI practices, data protection, and regulatory compliance.

Support and enablement functions include training coordinators, documentation specialists, and technical support personnel who facilitate knowledge sharing, assist project teams, and maintain organizational capabilities for effective agent development and operation.

Operational Focus

The CoE executes four primary operational activities that drive organizational AI maturity. Developing standards and guidelines creates structured frameworks for agent design, development, testing, deployment, and operation. These frameworks promote consistency and quality by integrating lessons from past projects and industry best practices.

Training and capability building delivers educational programs, workshops, and mentoring services that develop organizational expertise and ensure teams have access to current knowledge about AI technologies, development methodologies, and governance requirements.

Technology evaluation and platform management involves systematic assessment of available tools, frameworks, and platforms to identify optimal solutions while managing shared infrastructure and development environments that support efficient project execution.

Performance monitoring and optimization tracks key performance indicators across multiple agent implementations, identifies improvement opportunities, and facilitates sharing of successful optimization strategies that enhance overall organizational capabilities and return on investment.

Implementation

Successful AI agent deployment requires systematic execution that transforms strategic plans into operational reality. This methodology focuses on the tactical aspects of bringing agents from concept to production while maintaining quality, security, and performance standards.

Phase 1: Technical Foundation Setup

Development environment configuration establishes the infrastructure needed for agent development. Th environment includes version control systems, development pipelines, testing frameworks, and staging environments. These components support iterative development while ensuring code quality and reliable deployment.

Data pipeline architecture implements robust data collection, storage, and processing capabilities to ensure agents have consistent access to high-quality information. Effective implementation includes establishing data governance protocols, monitoring data quality, and automating data preparation workflows to support both development and production needs.

Security framework implementation deploys comprehensive security measures including access controls, encryption protocols, audit logging, and vulnerability scanning that protect sensitive data and system integrity throughout the development lifecycle and operational deployment.

Observability and monitoring infrastructure establishes comprehensive visibility into agent performance, system health, and operational metrics. This is achieved through centralized logging, performance metrics collection, real-time alerting, and distributed tracing capabilities. These systems enable proactive issue detection, performance optimization, and operational insights throughout development and production phases.

Integration infrastructure development creates the technical foundation for connecting agents with existing organizational systems, external APIs, and third-party services through standardized interfaces, middleware solutions, and error handling mechanisms that ensure reliable communication and data exchange.

Phase 2: Agent Development

Core agent logic development implements task execution and interaction capabilities that define agent behavior. This development involves translating functional requirements into executable code, implementing state management systems, and creating modular architectures that support maintainability and future enhancement.

AI model integration and optimization connects selected AI models with agent logic through appropriate APIs, SDKs, or embedded implementations while optimizing model performance for specific use cases.

Implementation includes implementing model versioning, A/B testing capabilities, and fallback mechanisms that ensure consistent agent performance.

Prompt engineering and response optimization develops, tests, and refines prompts for language model interactions to achieve desired agent responses while minimizing hallucinations, improving accuracy, and maintaining consistency across different usage scenarios.

User interface and experience implementation creates intuitive interfaces that enable effective human-agent interaction through web applications, mobile interfaces, chat systems, or API endpoints that support the intended user experience while maintaining accessibility and usability standards.

Phase 3: Testing and Validation

Automated testing involves creating comprehensive frameworks to validate agent functionality, performance, and reliability. This includes unit tests, integration tests, and end-to-end scenarios that simulate realistic usage patterns and edge cases.

Performance benchmarking and load testing evaluates agent performance under various conditions including normal operation, peak load scenarios, and stress conditions to ensure agents meet established performance criteria and can handle expected usage volumes.

Security testing and vulnerability assessment conducts systematic security evaluation including penetration testing, data protection validation, and compliance verification that confirms agents meet organizational and regulatory security requirements.

User acceptance testing coordination facilitates structured testing with actual users to validate that agents deliver intended functionality, provide positive user experiences, and meet practical usability requirements in real-world operational contexts.

Phase 4: Production Deployment

Deployment orchestration manages rollout strategies—such as blue-green deployments, canary releases, and phased rollouts—to minimize risk and enable rapid response to issues during initial production.

Monitoring and alerting system activation deploys comprehensive real-time monitoring that tracks agent performance, system health, resource utilization, and user interactions while providing immediate notification of issues requiring attention or intervention.

Implementing an operational support involves setting up help desk procedures, escalation protocols, and troubleshooting guides. It also includes maintaining detailed incident tracking and documentation to support accountability and continuous improvement.

Production data management implements data retention policies, backup procedures, and archival strategies that ensure business continuity while maintaining compliance with data protection regulations and organizational policies.

Phase 5: Continuous Optimization

Performance analytics involves systematically analyzing operational metrics, user behavior, and business impact indicators to uncover optimization opportunities and assess value delivery against defined objectives.

Iterative enhancement requires structured processes to refine agent capabilities based on performance analysis, user feedback, and shifting business needs. Improvements are introduced through controlled update cycles that safeguard system stability while driving continuous progress.

Model maintenance and evolution management creates systematic approaches for model retraining, fine-tuning, and updating that maintain agent performance while adapting to changing data patterns, user needs, and business requirements.

Knowledge capture ensures up-to-date documentation of agent capabilities, operational procedures, troubleshooting guides, and lessons learned. This supports effective day-to-day operations and lays the foundation for future development.

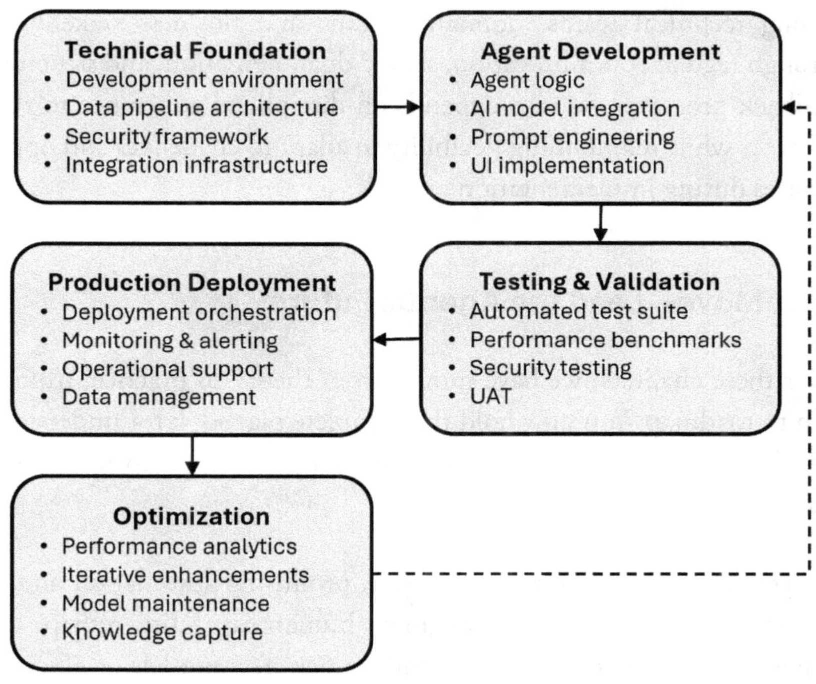

Figure 17: AI Agent Implementation Phases

Execution Principles for Success

The incremental delivery approach focuses on releasing functional agent capabilities early and often, allowing quick feedback, course correction, and visible value that builds organizational confidence. This methodology focuses on risk-aware implementation, maintaining thorough assessment and mitigation strategies throughout execution. It addresses technical risks, operational challenges, and business continuity factors that could affect successful deployment. Quality-first development ensures agents meet performance, reliability, and user experience standards through rigorous testing and code reviews before deployment.

Collaborative development practices facilitate effective collaboration among technical teams, domain experts, and business stakeholders through regular communication, shared documentation, and structured feedback processes. Success depends on disciplined adherence to these practices while maintaining flexibility to adapt to challenges and opportunities during implementation.

Your Move—Lead the Agentic Future

Over these chapters, we have moved from theory to practice, from vision to roadmap. You now hold the complete playbook for understanding, designing, deploying, and scaling AI agents that deliver real business value.

AI agents are today's competitive edge, promising automation and intelligent partnership that can augment human capability, reshape customer and employee experience, and unlock new models of efficiency and growth.

But owning the playbook isn't enough. **Now it's your move.**

As Alan Kay once said, "The best way to predict the future is to invent it."[2]

Begin with a manageable use case and build momentum from there. Unite your team behind a clear goal, choose the right use case, and build with purpose.

Measure success, scale what works, and build trust.

Above all, lead with vision. The organizations that succeed will be those whose leaders see AI agents not as experiments, but as a strategic imperative.

Your agent journey doesn't end here—it begins now.

ACKNOWLEDGEMENTS

I am filled with immense gratitude for the extraordinary support that made this book possible. *Leading With AI Agents* would not exist without the invaluable contributions of so many remarkable individuals who shared their wisdom, time, and encouragement throughout this journey.

First and foremost, I owe a profound debt of gratitude to my family. My heartfelt thanks go to my wife for her unwavering patience and encouragement during the countless hours spent researching, writing, and refining this work. I am eternally grateful to my two children for their boundless faith and steadfast support as I navigated the complexities of bringing this knowledge to life.

I am deeply indebted to my mentors—Scott Herren, Mark Templeton, Bill Siu, and Ken Robinson—whose exceptional leadership, humility, and wisdom continue to shape my thinking. Their encouragement to share knowledge and contribute to the broader community was instrumental in inspiring this book.

This work was greatly enriched by insights from industry leaders, Trevor Rodrigues-Templar, Tyler Jewell, Jim Goldfinger, and Deepak Dube, who generously shared their teams' experiences with AI agents. Their perspectives helped bridge the gap between theory and real-world application.

My sincere appreciation goes to the reviewers—especially Mariette Wharton, JeanAnn Nichols, Claus Moldt, Sam Beera, Marley Spector, and Murthy Garimella— whose thoughtful feedback on early drafts elevated the clarity and impact of this work.

The AI and technology community deserves special recognition. This book stands on the shoulders of giants—the researchers, engineers, and pioneers at organizations like Anthropic, Google, OpenAI, Hugging Face, and Microsoft, and countless startups whose work advances our collective understanding of AI agents. Their collaborative spirit and willingness to openly share knowledge, code, and insights are what make a book like this possible and propel the entire field forward.

This journey—from experimental AI to production-ready agents—has been a shared endeavor. I'm deeply honored to contribute to this transformation and grateful to all who continue to champion the positive potential of artificial intelligence.

REFERENCES

Introduction

1. Chui, Michael, et al. "The Economic Potential of Generative AI: The Next Productivity Frontier." *McKinsey & Company*, June 14, 2023. https://www.mckinsey.com/capabilities/mckinsey-digital/our-insights/the-economic-potential-of-generative-ai-the-next-productivity-frontier.
2. Brennan, Morgan. "Watch CNBC's Full Interview with Billionaire Investor Mark Cuban." *CNBC*, March 4, 2024. Video, 12:06. https://www.cnbc.com/video/2024/03/04/watch-cnbcs-full-interview-with-billionaire-investor-mark-cuban.html.

Chapter 1

1. Russell, Stuart, and Peter Norvig. *Artificial Intelligence: A Modern Approach.* 4th ed. Pearson, 2021.
2. SRI International. "Shakey the Robot." Accessed July 7, 2025. https://www.sri.com/pioneers/shakey-the-robot.
3. Brooks, Rodney A. "How to Build Complete Creatures Rather than Isolated Cognitive Simulators." Technical Report. Cambridge, MA: MIT Artificial Intelligence Laboratory, 1987. https://people.csail.mit.edu/brooks/papers/how-to-build.pdf
4. Rao, Anand S., and Michael P. Georgeff. "Modeling Rational Agents within a BDI-Architecture." In *Proceedings of the 2nd International Conference on Principles of Knowledge Representation and Reasoning (KR'91)*, 473–84. Cambridge, MA, April 1991.
5. Tampuu, Ardi, et al. "Multiagent Cooperation and Competition with Deep Reinforcement Learning." *arXiv*, November 28, 2015. https://arxiv.org/abs/1511.08779.
6. Xie, Junlin, et al. "Large Multimodal Agents: A Survey." *arXiv* February 23, 2024. https://arxiv.org/abs/2402.15116.
7. Zendesk. "What Are NLP Chatbots and How Do They Work?" *Zendesk*, August 5, 2024. https://www.zendesk.com/blog/nlp-chatbot.
8. Reda, Mohamed, et al. "Path Planning Algorithms in the Autonomous Driving System: A

Comprehensive Review." *Robotics and Autonomous Systems* 174 (2024): 104630. https://doi.org/10.1016/j.robot.2024.104630.

9. Acowebs. "How the Robots and Drones Delivery Could Transform Quick Commerce." *Acowebs*, October 21, 2024. https://acowebs.com/robots-drone-delivery/
10. Maleki Varnosfaderani, Shiva, and Mohamad Forouzanfar. "The Role of AI in Hospitals and Clinics: Transforming Healthcare in the Digital Era." *Bioengineering* 11, no. 4 (March 2024): 337. https://doi.org/10.3390/bioengineering11040337.
11. Simons, Alex. "Simons' Strategies: Renaissance Trading Unpacked." *LuxAlgo*, June 13, 2025. https://www.luxalgo.com/blog/simons-strategies-renaissance-trading-unpacked.
12. Royal, James, and Dayana Yochim. "Betterment vs. Wealthfront: Which Is Best for You?" *Bankrate*, June 11, 2025. https://www.bankrate.com/investing/betterment-vs-wealthfront.
13. Spiceworks. "Recommendation Engines: How Amazon and Netflix Are Winning the Personalization Battle." *Spiceworks*, June 28, 2016. https://www.spiceworks.com/marketing/customer-experience/articles/recommendation-engines-how-amazon-and-netflix-are-winning-the-personalization-battle.
14. Simon, Matt. "A Guide to the Bot-Eat-Bot World of Warehouse Robots." *Wired*, July 29, 2021. https://www.wired.com/story/a-guide-to-the-bot-eat-bot-world-of-warehouse-robots.
15. Dohmke, Thomas. "GitHub Copilot Is Generally Available to All Developers." *The GitHub Blog*, June 21, 2022. https://github.blog/2022-06-21-github-copilot-is-generally-available-to-all-developers.
16. Your Security Connection. "Smart Home Systems with Alexa and Google Assistant." *Your Security Connection*, 2025. https://www.yoursecurityconnection.com/smart-home-security-via-alexa-google
17. Tecknexus. "Rule-Based vs. LLM-Based AI Agents: A Side-by-Side Comparison." Last modified 2024. https://tecknexus.com/rule-based-vs-llm-based-ai-agents-a-side-by-side-comparison
18. GeeksforGeeks. "Model-Based Reflex Agents in AI." Last modified May 17, 2024. https://www.geeksforgeeks.org/artificial-intelligence/model-based-reflex-agents-in-ai.
19. IBM. "What Is AI Agent Planning?" Last modified March 24, 2025. https://www.ibm.com/think/topics/ai-agent-planning.
20. GeeksforGeeks. "Utility-Based Agents in AI." *GeeksforGeeks*. Last modified July 30, 2024. https://www.geeksforgeeks.org/artificial-intelligence/utility-based-agents-in-ai.
21. IBM. "What Is AI Agent Learning?" Last modified March 28, 2025. https://www.ibm.com/think/topics/ai-agent-learning.
22. IBM. "Types of AI Agents." *IBM Think*. Accessed July 12, 2025. https://www.ibm.com/think/topics/ai-agent-types.
23. Schluntz, Erik, and Barry Zhang. "Building Effective Agents." *Anthropic*, December 19, 2024. https://www.anthropic.com/engineering/building-effective-agents.
24. Model Context Protocol. "MCP: A Protocol for LLM App Interoperability." Accessed July 7, 2025. https://modelcontext.dev.

25. Scrapfly. "Guide to Understanding and Developing LLM Agents." February 19, 2025. https://scrapfly.io/blog/posts/practical-guide-to-llm-agents.
26. IBM. "What Is AI Agent Memory?" Last modified March 18, 2025. https://www.ibm.com/think/topics/ai-agent-memory.
27. Alake, Richmond. "Building Intelligent AI Agents with Memory: A Complete Guide." *DEV.to*, July 8, 2025. https://dev.to/bredmond1019/building-intelligent-ai-agents-with-memory-a-complete-guide-5gnk.
28. TiDB. "LangChain Memory Implementation: A Comprehensive Guide." December 13, 2024. https://www.pingcap.com/article/langchain-memory-implementation-a-comprehensive-guide.
29. Akka Documentation. "Introduction to Actors." November 6, 2019. https://doc.akka.io/libraries/akka-core/current/typed/actors.html.
30. IBM. "What Is Chain-of-Thought (CoT) Prompting?" Last modified April 23, 2025. https://www.ibm.com/think/topics/chain-of-thoughts.
31. Yao, Shunyu, Dian Yu, et al. "Tree of Thoughts: Deliberate Problem Solving with Large Language Models." *arXiv* May 17, 2023. https://arxiv.org/abs/2305.10601.
32. Ceurstemont, Sandrine. "Self-Correction in Large Language Models." *Communications of the ACM*, February 26, 2025. https://cacm.acm.org/news/self-correction-in-large-language-models.
33. Kim, Jiin, Byeongjun Shin, Jinha Chung, and Minsoo Rhu. "The Cost of Dynamic Reasoning: Demystifying AI Agents and Test-Time Scaling from an AI Infrastructure Perspective." *arXiv* June 9, 2025. https://arxiv.org/abs/2506.04301.
34. Anthropic. "Tool Use and Function Calling." Accessed July 9, 2025. https://www.anthropic.com/index/tool-use-and-function-calling.
35. Shingo, Shigeo. *Zero Quality Control: Source Inspection and the Poka-Yoke System*. Translated by Andrew P. Dillon. Portland, OR: Productivity Press, 1986.

Chapter 3

1. Chakraborty, Amrita, and Arpan Kumar Kar. "Swarm Intelligence: A Review of Algorithms." In *Nature-Inspired Computing and Optimization*, edited by Xin-She Yang, 475–94. Modeling and Optimization in Science and Technologies, vol. 10. Cham: Springer, 2017. https://link.springer.com/chapter/10.1007/978-3-319-50920-4_19.
2. Microsoft. "Overview of the Model Context Protocol in AI Agent Development." *Microsoft Learn*. June 20, 2024. https://learn.microsoft.com/en-us/Copilot/overview-model-context-protocol.
3. Google for Developers. "Agent2Agent Protocol." *Google for Developers*. Accessed July 7, 2025. https://developers.google.com/agents/a2a/overview.

4. Milvus. "How Do Agents Collaborate in a Multi-Agent System?" *Milvus*. Accessed July 23, 2025. https://milvus.io/ai-quick-reference/how-do-agents-collaborate-in-a-multiagent-system.
5. Milvus. "How Do Agents Collaborate in a Multi-Agent System?" *Milvus*. Accessed July 23, 2025. https://milvus.io/ai-quick-reference/how-do-agents-collaborate-in-a-multiagent-system.
6. Milvus. "How Do Agents Compete in a Multi-Agent System?" *Milvus*. Accessed July 23, 2025. https://milvus.io/ai-quick-reference/how-do-agents-compete-in-a-multiagent-system.
7. Smythos. "Multi-Agent Systems and Negotiation." *Smythos*. Accessed July 23, 2025. https://smythos.com/developers/agent-development/multi-agent-systems-and-negotiation.
8. Milvus. "What Are Hierarchical Multi-Agent Systems?" *Milvus*. Accessed July 23, 2025. https://milvus.io/ai-quick-reference/what-are-hierarchical-multiagent-systems.
9. Martínez-Gómez, M., E. Martínez-García, and F. de la Cruz-Merino. "Establishing a Relationship between Features and Process Variables for Monitoring a Hydrocyclone under Abnormal Conditions." *Computers & Industrial Engineering* 59, no. 6 (2010): 1076–86. https://doi.org/10.1016/j.cie.2010.07.013.
10. Supply Chain Digital. "Amazon's Bid to Revolutionise Warehouse Automation." *Supply Chain Digital*. Accessed July 23, 2025. https://supplychaindigital.com/digital-supply-chain/amazon-warehouse-automation-ai-revolution.
11. ThinkLabs AI. "Physics-Informed Agents Enable Smarter Energy Distribution in Modern Smart Grids." *ThinkLabs AI Whitepaper*. Accessed July 23, 2025. https://thinklabs.ai/case-studies/physics-informed-agents-smart-grids.
12. BytePlus. "Multi-Agent Systems Case Studies: Real-World Applications and Insights." *BytePlus*. Accessed July 23, 2025. https://www.byteplus.com/en/topic/400847.
13. Fraunhofer IFF. "Collaborating Robots in Assembly Processes (ColRobot)." *Fraunhofer IFF*. Accessed July 23, 2025. https://www.iff.fraunhofer.de/en/business-units/robotic-systems/colrobot.html.

Chapter 4

1. Oxipit. "Case Study: Embracing AI for Efficient Chest X-Ray Reporting and Quality Assurance." *Oxipit*. Accessed July 25, 2025. https://oxipit.ai/case-study/case-study-embracing-ai-for-efficient-chest-x-ray-reporting-and-quality-assurance.
2. Microsoft. "AI in Word—Features and Benefits | Microsoft 365 Copilot." *Microsoft*. Accessed July 9, 2025. https://www.microsoft.com/en-us/microsoft-365/word/word-ai.
3. Mallidi, Reddy. *AI Unleashed: A Leader's Playbook to Master AI for Business Excellence*. Manuscripts Press, 2024.
4. Miller, Gabby. "Breaking Down the Lawsuit Against Character.AI Over Teen's Suicide." *TechPolicy.Press*, October 23, 2024. https://www.techpolicy.press/breaking-down-the-lawsuit-against-characterai-over-teens-suicide.

Chapter 5

1. LangChain. "LangChain: Build Context-Aware Reasoning Applications." *GitHub.* Accessed July 9, 2025. https://github.com/langchain-ai/langchain.
2. Akka. "Additional Samples." *Akka Documentation.* Accessed July 9, 2025. https://doc.akka.io/getting-started/samples.html.
3. CrewAI. "Using Multimodal Agents." *CrewAI Documentation.* Accessed July 9, 2025. https://docs.crewai.com/en/learn/multimodal-agents.

Chapter 6

1. Jamali, Ruhollah, and Sanja Lazarova-Molnar. "On the Relationship between Model Complexity and Decision Support in Agent-Based Modeling and Simulation." *Simulation Notes Europe* 34, no. 3 (2024): 177–80. https://www.sne-journal.org/sne-volumes/volume-34/sne-343-articles/on-the-relationship-between-model-complexity-and-decision-support-in-agent-based-modeling-and-simulation.
2. Bronsdon, Conor. "How to Detect and Prevent Malicious Agent Behavior in Multi-Agent Systems." *Galileo*, April 8, 2025. https://galileo.ai/blog/malicious-behavior-in-multi-agent-systems.
3. Song, Yifan, Guoyin Wang, Sujian Li, and Bill Yuchen Lin. "The Good, The Bad, and The Greedy: Evaluation of LLMs Should Not Ignore Non-Determinism." *arXiv* July 15, 2024. https://arxiv.org/html/2407.10457.
4. Epperson, Will, Gagan Bansal, Victor Dibia, Adam Fourney, Jack Gerrits, Erkang Zhu, and Saleema Amershi. "Interactive Debugging and Steering of Multi-Agent AI Systems." *arXiv* March 3, 2025. https://arxiv.org/html/2503.02068.
5. Milvus. "What Is the Brittleness Problem in AI Reasoning?" *Milvus*, 2025. https://milvus.io/ai-quick-reference/what-is-the-brittleness-problem-in-ai-reasoning.
6. Park, Joseph, Lulu Ito, and Tim Burress. "Modularity in AI: Understanding the Building Blocks of Intelligence." *Digital Architecture Lab*, October 2, 2024. https://dalab.xyz/en/blog/modularity-in-ai-understanding-the-building-blocks-of-intelligence.
7. Bronsdon, Conor. "How to Test AI Agents + Metrics for Evaluation." *Galileo*, December 19, 2024. https://galileo.ai/blog/how-to-test-ai-agents-evaluation.
8. Google Cloud. "What Is Human-in-the-Loop (HITL) in AI & ML?" *Google Cloud*, 2025. https://cloud.google.com/discover/human-in-the-loop.
9. Duta, Grig. "What Is Prompt Management? Tools, Tips and Best Practices." *Qwak*, May 16, 2024. https://www.qwak.com/post/prompt-management.
10. Kittel, Chad, and Clayton Siemens. "AI Agent Orchestration Patterns." *Microsoft Learn*, July 18, 2025. https://learn.microsoft.com/en-us/azure/architecture/ai-ml/guide/ai-agent-design-patterns.

11. Kaur, Jagreet. "Streamlining Tool Integration with AI Agents." *Akira AI*, October 29, 2024. https://www.akira.ai/blog/streamlining-tool-integration-with-ai-agents.
12. Meegle. "Language Model Lifecycle Management." *Meegle*, July 10, 2025. https://www.meegle.com/en_us/topics/natural-language-processing/language-model-lifecycle-management.
13. Srivastava, Dhairya. "AI-Powered Test Automation in CI/CD Pipelines: A Complete Guide." *Quash*, March 31, 2025. https://quashbugs.com/blog/the-role-of-ci-cd-pipelines-in-ai-powered-test-automation.
14. Contentsquare. "10 A/B Testing Metrics + KPIs You Need to Track." *Contentsquare*, October 31, 2024. https://contentsquare.com/guides/ab-testing/metrics.
15. Shamim, Isha, and Rekha Singhal. "Methodology for Quality Assurance Testing of LLM-Based Multi-Agent Systems." In *Proceedings of the 4th International Conference on AI-ML Systems (AIMLSystems '24)*, Article 27, 1–5. March 5, 2025. https://dl.acm.org/doi/full/10.1145/3703412.3703439.
16. Jaeger Community. "Distributed Tracing Best Practices." *Jaeger Documentation*. Accessed August 1, 2025. https://www.jaegertracing.io/docs/latest/best-practices.
17. LangChain AI. "LangSmith Documentation: Tracing and Monitoring." *LangSmith User Guide*. Accessed August 1, 2025. https://docs.smith.langchain.com/tracing.
18. Fowler, Martin. "Structured Logging." *Martin Fowler's Blog*, May 15, 2023. https://martinfowler.com/articles/structured-logging.html.
19. Chen, Li, et al. "Anomaly Detection in Distributed Systems: A Survey." *ACM Computing Surveys* 54, no. 3 (2021): 1–35. https://doi.org/10.1145/3447556.

Chapter 7

1. Master of Code Global. "AI in Customer Service Statistics." *Master of Code Global.* Accessed August 3, 2025. https://masterofcode.com/blog/ai-in-customer-service-statistics.
2. Interface.ai. . "interface.ai Unveils Q1 2025 Release: Transforming Banking AI with Next-Level Automation, Security, and Personalization." *Yahoo! Finance*, April 1, 2025. https://finance.yahoo.com/news/interface-ai-unveils-q1-2025-123000133.html.
3. Samal, Ansuman. "Enhancing Process Automation with AI: The Role of Intelligent Automation in Business Efficiency." April 30, 2022. Preprint posted on ResearchGate. https://www.researchgate.net/publication/385163023_Enhancing_process_automation_with_AI_The_role_of_intelligent_automation_in_business_efficiency.
4. EY. "Case Study: How AI Automated Insurance Claims." *EY.* Accessed August 3, 2025. https://www.ey.com/en_us/insights/financial-services/emeia/how-a-nordic-insurance-company-automated-claims-processing.
5. The CDO TIMES. "Case Study: Amazon's AI-Driven Supply Chain: A Blueprint for the Future of Global Logistics." *The CDO TIMES*, August 23, 2024. https://cdotimes.com/2024/08/23/case-study-amazons-ai-driven-supply-chain-a-blueprint-for-the-future-of-global-logistics.

6. UPS. "UPS To Enhance ORION With Continuous Delivery Route Optimization." Press Release, January 29, 2020. https://about.ups.com/us/en/newsroom/press-releases/innovation-driven/ups-to-enhance-orion-with-continuous-delivery-route-optimization.html.
7. Kyros AML. "The Role of AI and Machine Learning in KYC and AML Compliance." *Kyros AML.* May 20, 2023. https://kyrosaml.com/the-role-of-ai-and-machine-learning-in-kyc-and-aml-compliance.
8. HashStudioz. "AI Transforming Financial Planning: The Role of Robo-Advisors." *HashStudioz.* December 26, 2024. https://www.hashstudioz.com/blog/how-ai-is-transforming-financial-planning-through-robo-advisors.
9. Campbell, Dakin. "JPMorgan Chase Is Rolling Out an AI Assistant for All Its Wealth Advisers." *Business Insider*, October 19, 2023. https://www.businessinsider.com/jpmorgan-chase-ai-assistant-wealth-advisers-coach-ai-2023-10.
10. Singh, Gurjot, Prabhjot Singh, and Maninder Singh. "Advanced Real-Time Fraud Detection Using RAG-Based LLMs." *arXiv* preprint arXiv:2501.15290, January 2025. https://arxiv.org/pdf/2501.15290.
11. Kody Technolab. "AI in Stock Trading: Insights for CEOs to Drive Growth and Profit." *Kody Technolab*, May 21, 2025. https://kodytechnolab.com/blog/ai-in-stock-trading.
12. Spectral AI. "Artificial Intelligence in Medical Diagnosis: Medical Diagnostics and AI." *Spectral AI*, October 4, 2024. https://www.spectral-ai.com/blog/artificial-intelligence-in-medical-diagnosis-how-medical-diagnostics-are-improving-through-ai.
13. Tariq, M, et al. "The Role of AI in Hospitals and Clinics: Transforming Healthcare in the 21st Century." *Cureus* 16, no. 3 (March 2024): e56561. https://pmc.ncbi.nlm.nih.gov/articles/PMC11047988.
14. Harvard Gazette. "New AI Tool Can Diagnose Cancer, Guide Treatment, Predict Patient Survival." *Harvard Gazette*, October 9, 2024. https://news.harvard.edu/gazette/story/2024/09/new-ai-tool-can-diagnose-cancer-guide-treatment-predict-patient-survival.
15. Varsha, G., S. K. Singh, and R. K. Singh. "Artificial Intelligence (AI) Applications in Drug Discovery and Drug Delivery: Revolutionizing Personalized Medicine." *International Journal of Molecular Sciences* 25, no. 22 (November 2024): 13628. https://pmc.ncbi.nlm.nih.gov/articles/PMC11510778.
16. G., Srushti. "AI-Driven Personalized Medicine and Drug Discovery." March 1, 2024. Preprint posted on ResearchGate. https://www.researchgate.net/publication/389504711_AI-Driven_Personalized_Medicine_and_Drug_Discovery.
17. Aviso. "Revenue Forecasting." *Aviso.* Accessed July 9, 2025. https://www.aviso.com/blog/revenue-forecasting.
18. Akka.io. . "Personalized User Experiences Drive Increased Advertising Revenue at Tubi." *Akka.* Accessed August 14, 2025. https://akka.io/customer-stories/personalized-user-experiences-drive-inc.

19. Amazon. "How Amazon Is Using Generative AI to Improve Product Recommendations and Descriptions." *About Amazon*, September 19, 2024. https://www.aboutamazon.com/news/retail/amazon-generative-ai-product-search-results-and-descriptions.
20. Amazon. "Amazon Deploys over 1 Million Robots and Launches New AI Foundation Model." *About Amazon*, June 30, 2025. https://www.aboutamazon.com/news/operations/amazon-million-robots-ai-foundation-model.
21. Microsoft. "Introducing Microsoft 365 Copilot — Your Copilot for Work." *Microsoft News Center*, March 16, 2023. https://news.microsoft.com/reinventing-productivity.
22. Tesla. "The Bigger Picture on Autopilot Safety." *Tesla Blog*, January 9, 2024. https://www.tesla.com/blog/bigger-picture-autopilot-safety.
23. Tesla. "Tesla Reveals FSD Beta Accident Rate for First Time; Compares It to Autopilot and National Average." *Not a Tesla App*, April 25, 2023. https://www.notateslaapp.com/news/1370/tesla-reveals-fsd-beta-accident-rate-and-compares-it-to-autopilot-and-national-average.
24. Aalpha Information Systems. "Agent as a Service (AaaS): A Comprehensive Guide." *Aalpha.net*, May 12, 2025. https://www.aalpha.net/blog/agent-as-a-service-aaas-comprehensive-guide.
25. IBM. "What Is Hyper-personalization?" *IBM.com*, January 31, 2025. https://www.ibm.com/think/topics/hyper-personalization.
26. Klover AI. "How Autonomous AI Agents Are Transforming Public Service Delivery." *Klover.ai*, May 6, 2025. https://www.klover.ai/how-autonomous-ai-agents-are-transforming-public-service-delivery.
27. Neuralt. "How Predictive Analytics Can Optimize Revenue Growth." *Neuralt.com*, February 24, 2025. https://www.neuralt.com/news-insights/how-predictive-analytics-can-optimize-revenue-growth.
28. Mallidi, Reddy. *AI Unleashed: A Leader's Playbook to Master AI for Business Excellence*. Manuscripts Press, 2024.
29. Inero Software. "LLM Implementation and Maintenance Costs for Businesses: A Detailed Breakdown." *Inero Software*. Last modified May 16, 2025. https://inero-software.com/llm-implementation-and-maintenance-costs-for-businesses-a-detailed-breakdown.

Chapter 8

1. Pingdom. "Page Load Time vs. Response Time – What Is the Difference?" *Pingdom*. Last updated February 28, 2024. https://www.pingdom.com/blog/page-load-time-vs-response-time-what-is-the-difference/.
2. Kamble, Sangharshjit. "More Response/Processing Time in Document Intelligence Service." *Microsoft Q&A*. Last modified August 16, 2024. https://learn.microsoft.com/en-us/answers/questions/1856853/more-response-processing-time-in-document-intellig.

3. Hossain, Maruf. "How the Cost of Large Language Models Varies Across Different Industries and Use Cases." *Dev.to*. Last modified 2024. https://dev.to/marufhossain/how-the-cost-of-large-language-models-varies-across-different-industries-and-use-cases-1b0b.
4. LogicMonitor. "Uptime vs. Availability." *LogicMonitor*. Last modified 2024. https://www.logicmonitor.com/blog/uptime-vs-availability.
5. Li, Yongjie, et al. "Legal Document Summarization: Enhancing Judicial Efficiency through Automation Detection." *arXiv*, July 30, 2025. https://arxiv.org/html/2507.18952v1.
6. SayOne. "How Faster Responses Boosted an E-Commerce Store's Conversion Rates by 31%." *SayOne Tech Blog*. 2024. https://www.sayonetech.com/blog/how-faster-responses-boosted-an-e-commerce-stores-conversion-rates-by-31/.
7. Peter, Anishka, et al. "Multi-Agent Translation Team (MATT): Enhancing Low-Resource Language Translation through Multi-Agent Workflow." *SMU Data Science Review* 8, no. 3 (Winter 2024): Article 3. https://scholar.smu.edu/cgi/viewcontent.cgi?article=1288&context=datasciencereview.
8. Multimodal.dev. "23 AI Agent Performance Metrics for Leaders." *Multimodal.dev*. Last modified 2024. https://www.multimodal.dev/post/ai-agent-performance-metrics-for-leaders.
9. CustomerGauge. "NPS Healthcare Guide: 25 Healthcare NPS Benchmarks & Industry Guide." *CustomerGauge*. Last modified 2024. https://customergauge.com/benchmarks/blog/nps-healthcare-net-promoter-score-benchmarks.
10. Zendesk. "22 Customer Service Metrics to Measure Your Team's Performance." *Zendesk*. Last modified 2024. https://www.zendesk.com/blog/customer-service-metrics/.
11. Woebot Health. "Introduction to Woebot Health." *Woebot Health*. Accessed August 8, 2025. https://woebothealth.com/how-woebot-works/health/.
12. McKinsey & Company. "How Retailers Use Personalization to Drive Growth." *McKinsey & Company*. Last modified 2023. https://www.mckinsey.com/industries/retail/our-insights/personalizing-the-customer-experience-driving-differentiation-in-retail.
13. GotBilled. "Manual vs Automated Invoice Processing: A Cost Comparison." *GotBilled*, March 11, 2025. https://www.gotbilled.com/blog/manual-vs-automated-invoice-processing-a-cost-comparison.
14. American Customer Satisfaction Index (ACSI). "Telecommunications Study 2022–2023." Press Release. June 6, 2023. https://theacsi.org/news-and-resources/press-releases/2023/06/06/press-release-telecommunications-study-2022-2023/.
15. McKinsey & Company. "How Retailers Can Keep Up with Consumers." *McKinsey & Company*, October 2019. https://www.mckinsey.com/industries/retail/our-insights/how-retailers-can-keep-up-with-consumers.
16. Liu, Xing, et al. "Accuracy and Effects of Clinical Decision Support Systems on Clinical Practice: A Retrospective Observational Study." *Journal of Medical Internet Research* 22, no. 1 (2020). https://pmc.ncbi.nlm.nih.gov/articles/PMC6997922/.

17. Zendesk. "What Is First Contact Resolution (FCR)? Benefits + Best Practices." *Zendesk*, May 30, 2017. https://www.zendesk.com/blog/first-contact-resolution-friend-foe-frenemy/.
18. Papers With Code."ImageNet Benchmark (Image Classification)." *Papers With Code*. Accessed May 26, 2025.
https://paperswithcode.com/sota/image-classification-on-imagenet?metric=Top+5+Accuracy.
19. Borji, Ali. "Breaking Beyond COCO Object Detection." *OpenReview*, February 1, 2023. https://openreview.net/forum?id=hj7uBF92qvm.
20. Metaculus. "What Will the State-of-the-Art Performance on SuperGLUE Be on 2021-06-14?" *Metaculus*, June 14, 2021. https://www.metaculus.com/questions/5937/sota-on-superglue-on-2021-06-14/?invite=tLxPdB.
21. Dinan, Emily, et al. "The Second Conversational Intelligence Challenge (ConvAI2)." *arXiv*, January 31, 2019. https://arxiv.org/pdf/1902.00098.pdf.
22. AI Multiple. "AI in Healthcare: Challenges & Best Practices in 2025." *AI Multiple*, July 17, 2025. https://research.aimultiple.com/healthcare-ai/.
23. Scale AI. "AI-Powered Contract Data Extraction." *Scale AI*, August 2, 2023. https://scale.com/enterprise/prebuilt-applications/ai-powered-contract-data-extraction..
24. Synpulse. "Making Onboarding Invisible." *Synpulse*, July 12, 2023. https://www.synpulse.com/en/insights/making-onboarding-invisible.
25. Superagi. "Future of Inventory: How AI Agents Are Revolutionizing Demand Forecasting and Stock Optimization in 2025." *Superagi*, June 30, 2025. https://superagi.com/future-of-inventory-how-ai-agents-are-revolutionizing-demand-forecasting-and-stock-optimization-in-2025/.

Chapter 9

1. IBM. *"OAuth 2.0 and OIDC Support." IBM Documentation*. Accessed July 18, 2025. https://www.ibm.com/docs/en/sva/11.0.0?topic=configuration-oauth-20-oidc-support
2. OpenID Foundation. "*How OpenID Connect Works.*" *OpenID Foundation*. Accessed July 18, 2025. https://openid.net/foundation/how-connect-works/.
3. Google Cloud. "Mutual TLS Overview | Load Balancing. " *Google Cloud*. Accessed July 18, 2025. https://cloud.google.com/load-balancing/docs/https/setting-up-mtls-ca-ccm.
4. European Union. *Regulation (EU) 2016/679 of the European Parliament and of the Council of 27 April 2016 on the Protection of Natural Persons with Regard to the Processing of Personal Data and on the Free Movement of Such Data (General Data Protection Regulation). Official Journal of the European Union*, April 27, 2016. https://eur-lex.europa.eu/eli/reg/2016/679/oj.
5. California Privacy Protection Agency. *California Consumer Privacy Act Regulations*. Effective January 2, 2024. https://cppa.ca.gov/regulations/pdf/cppa_regs.pdf.

6. DAMA International. *The DAMA Guide to the Data Management Body of Knowledge (DAMA-DMBOK2R).* 2nd ed., revised. Sedona, AZ: Technics Publications, LLC, 2024. https://www.dama.org/dmbok2r-infographics/.
7. European Parliament. "EU AI Act: First Regulation on Artificial Intelligence." European Parliament *News*, March 13, 2024. https://www.europarl.europa.eu/news/en/press-room/20240308IPR19015/eu-ai-act-first-regulation-on-artificial-intelligence.
8. National Institute of Standards and Technology. *Artificial Intelligence Risk Management Framework (AI RMF 1.0).* January 26, 2023. https://doi.org/10.6028/NIST.AI.100-1.
9. OECD. *AI Principles.* Paris: Organisation for Economic Co-operation and Development, 2019. https://www.oecd.org/en/topics/ai-principles.html.
10. International Organization for Standardization. *ISO/IEC 23053:2022 – Framework for Artificial Intelligence (AI) Systems Using Machine Learning (ML).* Geneva: ISO, 2022. https://www.iso.org/standard/74438.html..
11. International Organization for Standardization. *ISO/IEC 23894:2023 – Information Technology – Artificial Intelligence – Guidance on Risk Management.* Geneva: ISO, 2023. https://www.iso.org/standard/77304.html.

Chapter 10

1. Mallidi, Reddy. *How to Choose the Right AI Use Cases: Your Step-by-Step Workbook.* Independently published, August 30, 2024.
2. Kay, Alan. "Speaker Profile." *TED.* Accessed July 7, 2025. https://www.ted.com/speakers/alan_kay.

ABOUT THE AUTHOR

REDDY MALLIDI is Chief AI Officer and COO at J&R Consulting, where he leads AI strategy, implementation, and governance for Fortune 2000 companies. Trusted by C-suite leaders and boards, he helps organizations turn AI from hype into operational results.

Over the past few years, he has generated more than $150 million in annual savings by implementing large-scale AI, machine learning, and generative AI solutions. His technical journey began with neural networks in graduate school, and he has since led enterprise AI transformation across multiple industries.

Notable projects include automating export compliance screening for over 50 million accounts, reducing manufacturing waste by 55% through AI-driven SKU optimization, achieving over 95% accuracy in pharmaceutical quality inspections, and reconciling more than 80 million daily financial transactions using AI.

Beyond his technical leadership, Reddy has spearheaded business transformation initiatives—delivering double-digit profitability improvements at ADP, generating over $1 billion in incremental revenue and cost efficiencies at Intel, and scaling Autodesk's operations to support a twofold increase in revenue while achieving 95% customer satisfaction.

He also serves as a strategic board advisor to AI startups and is the author of *AI Unleashed: A Leader's Playbook to Master AI for Business Excellence* and *How to Choose the Right AI Use Cases*. He is a sought-after speaker at industry conferences and universities.

Reddy holds an MBA from Columbia University, an MS in Computer Science from Oklahoma State University, and an MTech in Civil Engineering from the Indian Institute of Technology, Kharagpur.

Growing up in rural India with limited educational access, he developed a deep belief in human potential—a conviction that now drives his vision of AI as an empowering force for positive change in business and society.

www.ingramcontent.com/pod-product-compliance
Lightning Source LLC
Chambersburg PA
CBHW071551210326
41597CB00019B/3200

* 9 7 9 8 9 9 3 1 0 2 9 0 0 *